INTRODUCTION TO THE STUDY OF ANIMAL POPULATIONS

INTRODUCTION TO THE STUDY OF ANIMAL POPULATIONS

H. G. Andrewartha

PHOENIX BOOKS

THE UNIVERSITY OF CHICAGO PRESS
CHICAGO & LONDON

PHOENIX SCIENCE SERIES

This book is also available in a clothbound edition from
THE UNIVERSITY OF CHICAGO PRESS

THE UNIVERSITY OF CHICAGO PRESS, CHICAGO & LONDON
The University of Toronto Press, Toronto 5, Canada

Published 1961. First Phoenix edition 1963. Third impression 1965. Printed in the United States of America

Contents

PART I: THEORY

Illustrations

Preface

My interest in animal ecology was first aroused by reading Elton's *Animal Ecology*. His definition of the scope of ecology which I quote in section 1.0 is still the best that I have met.

Notwithstanding the breadth of Elton's original definition the tendency of modern theoretical ecology has been to stress the interrelationships between animals at the expense of other components of environment which may also influence the 'distribution and numbers of animals in nature'. On the other hand the published works of practising ecologists, in such fields as economic entomology or the conservation of wildlife, have provided a wealth of information about how the distribution and numbers of animals in nature are in fact determined. Andrewartha and Birch attempted to systematize this information in *The Distribution and Abundance of Animals*.

The course in animal ecology that I teach in Adelaide makes use of Andrewartha and Birch's theory of environment. But I have felt the need for a compact text that the students may use. Also when the theory of animal ecology is taught this way it seems natural to join it to a practical course that is quantitative and largely experimental. In writing this book I have had these two objectives in mind. Part I gives a compact exposition of Andrewartha and Birch's theory of environment; Part II is a manual of practical exercises that arise from and are related to this theory.

Section 9.3 is addressed primarily to teachers and advanced students. If ecology is to be taught narrowly with the emphasis on the relationships between the animals in communities the issue of 'density-dependent factors' may not have to be faced seriously. But if ecology is to be taught broadly giving proper emphasis to all aspects of environment then the teacher must decide for himself what part, if any, the concept of density-dependent factors may play in a general theory of environment. I hope that section 9.3 may help him to reach a critical decision. The undergraduate should read section 9.33 as an introduction to Part II even if he does not read the rest of section 9.3 as an extension of Part I.

The experiments which are described in Part II may be divided into those dealing with methods for measuring dispersal, distribution and density and those dealing with physiological and behavioural responses of animals to particular components of environment such as temperature, food, other animals, etc. The theoretical background for the experiments in Part II is discussed in Part I, in relation to its proper place in the general theory. This arrangement has allowed me to set out the experiments in Part II with economy, as in a laboratory notebook; with respect to each group of experiments I have indicated the sections in Part I that are relevant; and then each experiment is set out under stereotyped headings viz. – (*a*) Purpose, (*b*) Material, (*c*) Apparatus, (*d*) Method, (*e*) Analysis of results. This arrangement also has the advantage of weaving theory and practice closely together.

Ecology, by its very nature, deals with populations of living animals and it becomes a dull subject unless it can be related to natural populations. Ideally the methods of estimating density, dispersal and distribution should be taught with populations that are living naturally in the field. I do this so far as it is practicable. But, in some instances the methods can also be taught by means of 'laboratory models', as, for example, in experiment 10.25 where we estimate the density of a population of *Tribolium* by the method of capture, marking, release and recapture. The laboratory model has the advantage that it can be done indoors during bad weather or during winter. Also it can usually be relied upon to give results that can be analysed.

Most of the animals that are available for study and the sorts of places where they live are characteristic of particular climatic or zoogeographic regions. This poses the problem of how to write a manual of practical animal ecology that might be useful outside the region where it was compiled. But ecological principles and techniques can be demonstrated with a wide variety of experimental animals, and I have chosen species that are cosmopolitan, or at least, are similar to those that are distributed widely in other continents, e.g. grain beetles, white cabbage butterfly, the garden snail and the house mouse. But, apart from these, there must be, in any region, many local species that are suitable for quantitative ecological experiments.

I have been helped not only to think but also to write more clearly by many long discussions with Dr T. O. Browning. My wife has read and criticized the text, and helped with the preparation of indexes. Many of the illustrations have been taken from *The Distribution and Abundance of Animals*, but a number are original. They were drawn by

Mrs P. E. Madge. I am grateful to Mr G. Wilkinson and Dr A. James for help with statistics, especially with the method for estimating dispersal.

H. G. A.

Adelaide 1959

PART I

Theory

CHAPTER I

The History and Scope of Ecology

1.0 HISTORICAL INTRODUCTION: THE APPROACH TO 'COMMUNITY ECOLOGY' THROUGH NATURAL HISTORY? OR THROUGH 'POPULATION ECOLOGY'?

In their *Principles of Animal Ecology* Allee *et al.* (1949) wrote a long introductory chapter on the history of the subject. They went back to the writings of Theophrastus, who was a friend of Aristotle, and who was interested in the sorts of animals and plants that could usually be found living together in 'communities'. Allee *et al.* also found much that was relevant to ecology in the writings of Pliny and other Romans, and, of course, in the writings of scholars who lived in various parts of Europe after the Renaissance. In more recent times the great French naturalists Reaumur (1663–1757) and Buffon (1707–88) and the German, Ernst Haeckel, who coined the word 'Oekologie' in 1869, all helped to build the foundations of the modern science.

The publication in 1927 of Elton's book *Animal Ecology* greatly influenced the development of modern ecology. The fashion in biology had previously been set by the publication in 1859 of *The Origin of Species* and the great controversies which followed. Taxonomy, comparative anatomy, evolution and phylogeny had been the chief interest of biologists for more than half a century. Elton, protesting against this perspective, wrote:

> In solving ecological problems we are concerned with *what animals do* in their capacity as whole, living animals, not as dead animals or as a series of parts of animals. We have next to study the circumstances under which they do these things, and, most important of all, the limiting factors which prevent them from doing certain other things. By solving these questions it is possible to discover the reasons for the *distribution and numbers of animals in nature.*

Elton's great contribution was to point out the importance of studying the distribution and abundance of animals in nature and to demarcate this as the proper field for ecology. He went on to say that the best way to go about this was to study the relationships between animals living together in 'communities'. He thought of a community as the complex of animals that are usually found living together in a 'habitat'. The meaning of 'habitat' was made clear in Elton's (1949) essay on 'population interspersion'. A habitat is defined as an area that seems to possess a certain uniformity with respect to physiography, vegetation or some other quality that the ecologist decides is important (or easily recognized). He decides on the limits of the habitat arbitrarily, and in advance, as a first step towards his study of the community.

The sort of relationships which Elton emphasized may be defined by explaining the concept of 'niche'. Elton (1927) said that all communities of plants and animals have a similar ground-plan; they all have their herbivores, carnivores and scavengers. In a wood there may be certain caterpillars which eat the leaves of trees, foxes which hunt rabbits and mice, beetles which eat springtails and so on. 'It is convenient to have some term to describe the status of an animal in its community, to indicate what it is *doing* and not merely what it looks like, and the term used is "niche" . . . The "niche" of an animal means its place in the biotic environment, *its relations to food and enemies.*' Thus the caterpillar and the mouse broadly occupy similar niches because they eat plants; the fox and the ladybird also fit broadly into similar niches because they both eat other animals in the community. But we need to have more information than this about 'niches'. For example, with respect to the 'niche' of the fox, it is not sufficient merely to know that 'foxes eat rabbits' and 'foxes can flourish on a diet of rabbit'. In addition we need to inquire 'how many foxes are there?' and 'what are their chances of getting enough food?'

So we may conveniently recognize three levels of complexity in the laws of ecology. Firstly, there are the laws governing the physiology and behaviour of *individuals* in relation to their environments. Secondly, there are the laws governing the numbers of animals in relation to the areas that they inhabit: this is sometimes called *population* ecology. And thirdly, there are the laws governing *communities*, which may be thought of as groups of interacting populations.

It seems reasonable to proceed upwards through these three levels of complexity – that is, to approach the study of populations through physiology and behaviour, and to approach the study of communities through the ecology of populations. That is what I have done in this book.

At present we are able to make satisfactory progress through the first two levels of complexity because we have a good basis of physiology and behaviour on which to build, and we have also accumulated a lot of information about the distribution and abundance of particular species. This arises from the work that has been done in pest control, conservation of wildlife, animal husbandry and so on. These investigations have had as their immediate purpose the discovery of methods for decreasing the abundance or restricting the distribution of pests, or for increasing the abundance or extending the distribution of species that are useful either because they are predators of pests, or are game or fur-bearers or are useful in some other way. In this literature Andrewartha and Birch (1954) found a wealth of facts and ideas that were relevant to a theory of ecology in which the idea of *population* was central.

Population ecology takes into account (amongst other things) the relationships of animals to their food, and to other sorts of animals that eat the same sort of food, or prey on them or are related to them in some other way. Consequently population ecology provides the means for studying the laws of 'niches' at the most fruitful level – that is at the level where populations interact. And it seems likely that a science of community ecology may eventually be built on the base provided by population ecology (see also sec. 6.4). But this is a very different matter from studying the 'community' that is circumscribed by the arbitrary definition of a 'habitat'. Elton's prediction that the laws governing the distribution and numbers of animals in nature would emerge from the study of the 'communities' that occur in 'habitats' has not been borne out by experience. The converse is more nearly true – that we must first discover the laws governing the distribution and abundance of animals before we can advance very far in our understanding of the relationships between the populations that make up a community. This is not to say that the study of the 'communities' that happen to occur in 'habitats' may not be continued for its own sake.

1.1 POPULATION ECOLOGY; THE STUDY OF THE DISTRIBUTION AND ABUNDANCE OF ANIMALS

Some time ago I spent a number of years studying the ecology of *Thrips imaginis* which is a small winged insect that lives and breeds in the flowers of a wide variety of garden plants, weeds, fruit trees and native shrubs around Adelaide, South Australia. One reason for choosing to work on this particular species was that it had recently (during the spring of 1931) multiplied to such enormous numbers that it had caused what was virtually a complete failure of the apple crops in the states of South Australia, Victoria and New South Wales. The apple-growing districts in South Australia are separated from those in Victoria by about 500 miles and those in New South Wales are still farther away. No one counted the number of *Thrips imaginis* in apple orchards very thoroughly that year. But we know now that it takes about 40 thrips to destroy one blossom. Starting with this figure, and taking into account the number of flowers that an apple tree might bear, I have guessed that there might have been about 20 million thrips per acre in most of the apple orchards in south-eastern Australia in 1931; they were also present in comparable numbers outside the orchards in gardens, fields and elsewhere.

We set out to explain the unusual abundance of *Thrips imaginis* during 1931; of course this meant that we had to try to explain the abundance of *Thrips imaginis* at all times – not merely during such extraordinary plagues. It seemed reasonable to start with the hypothesis that the plague had been caused primarily by weather because we could think of no other component of environment that would operate over such a vast area. So we set out to study all the ways that weather might influence the fecundity, speed of development and length of life – and hence the abundance – of *Thrips imaginis*. From the beginning it was clear that we would have nothing to explain unless we knew how many *Thrips imaginis* were present in our study-area. So we began counting the thrips in the large garden where we were working, and we continued to count them every day for 14 years. It was these counts that formed the fundamental basis around which we planned all our concomitant studies of behaviour, physiology, meteorology, etc. – in a word the ecology of *Thrips imaginis*.

A few years later I began to study the ecology of a grasshopper, *Austroicetes cruciata*, which, at that time, was causing a lot of damage to pastures and crops of wheat in one part of South Australia. During the next four or five years I found that swarms of this grasshopper occurred each year in a large but nevertheless well-defined zone (Fig. 1.01). I studied the old records and found that during the preceding 50 years there had been about seven outbreaks, and on every occasion the grasshoppers seemed to have been restricted to the same area that I had found them in during 1935–40. In this instance the most interesting scientific problem seemed to be to find an explanation for the *distribution* of this grasshopper, and to discover the reasons for the sharply defined boundaries to the area in which it was able to multiply to plague numbers. We discovered that the boundaries of the distribution were closely related to climate (Andrewartha, 1944). And it soon became clear that the components of climate which determined the limits of the distribution were the same as the components of weather that determined the abundance of the grasshoppers inside the area of

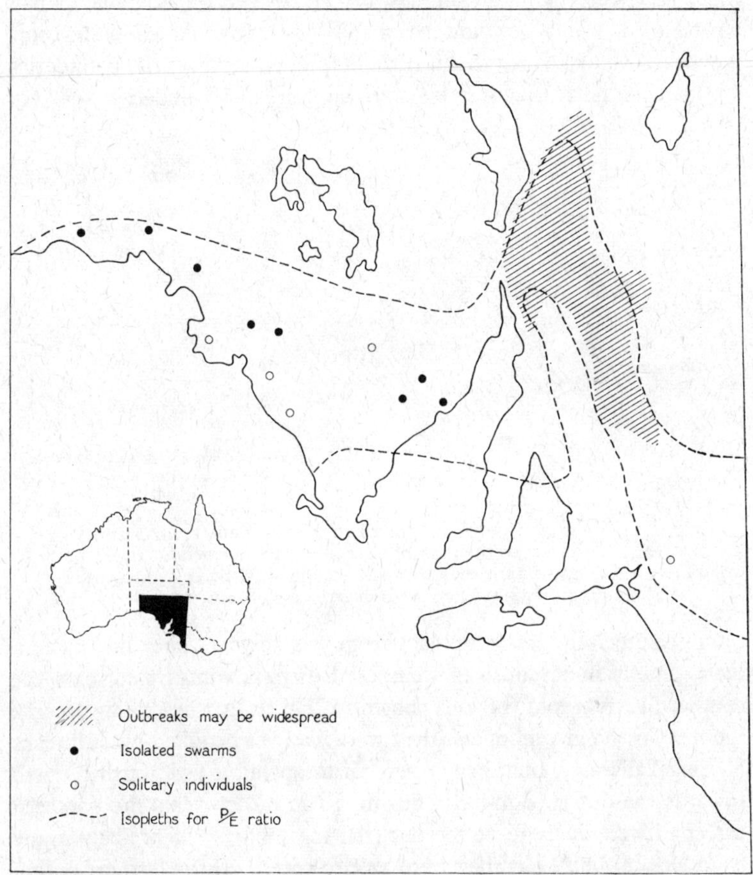

Fig. 1.01. The area in South Australia where *Austroicetes cruciata* may maintain a dense population when the weather is favourable. The isopleths for P/E ratio define the boundaries of a climatic zone to which the 'grasshopper belt' is largely confined. (After Andrewartha, 1944)

their distribution. In other words, there were not two separate problems of distribution and abundance to be explained, but only two aspects of the same problem. As our studies of *Thrips imaginis* progressed we saw that the same was true for this species too – as I have explained elsewhere (Andrewartha and Birch, 1954, p. 582).

In order to generalize this point I have drawn Figures 1.02 and 1.03. They are based on the observed distribution of *A. cruciata* but they also serve to illustrate the relationship between distribution and abundance that holds generally for all animals. For *A. cruciata* we were able to map a zone inside which we could reasonably expect to find the grasshoppers whenever we looked for them. Although we could draw the boundaries of this zone fairly precisely for any one year, the boundaries were not

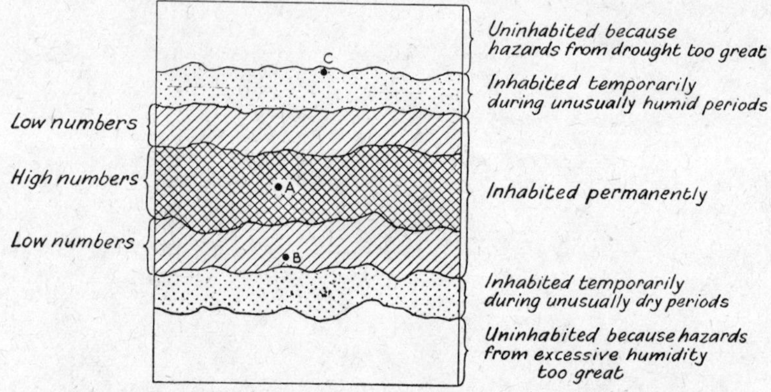

Fig. 1.02. Hypothetical map of the distribution of any animal based on the real distribution of the grasshopper *Austroicetes cruciata* (Fig. 1.01). For further explanation see text. The localities A, B, C also feature in Figure 1.03. (After Andrewartha and Birch, 1954)

found in precisely the same place from year to year. They fluctuated a little, chiefly in response to weather. During a run of wet years the southern extremes of the area shown in Figure 1.01 tended to become too wet for the grasshoppers (they were likely to be killed by outbreaks of fungal disease), but they were able to spread a little farther north towards the desert. Similarly, during a run of dry years the northern extremes were likely to be too dry (the grasshoppers were likely to die from lack of food), but they were able to spread a little farther south.

This is the sequence of events that I have tried to summarize in Figures 1.02 and 1.03. In Figure 1.02 the emphasis is on the fluctuating shape and size of the distribution; in Figure 1.03 I have emphasized fluctuations in abundance, not only from year to year but also from place to place; the two diagrams are linked by the localities A, B and C which are shown in both. The grasshoppers were usually more abundant in the central more favourable parts of the distribution (locality A) where the weather was usually neither too wet nor too dry. They were less abundant towards the margins of the distribution where the weather was more often too wet or too dry. The weather was never quite wet enough for long enough at B to reduce the population to zero; at C the weather was, from time to time, dry enough to wipe out the entire population; but during runs of wet years the area might be re-colonized temporarily. A little farther north the climate was too dry to permit even temporary colonization. Thus I found that when I had explained the abundance of the grasshoppers in all the places that they might occur I had also

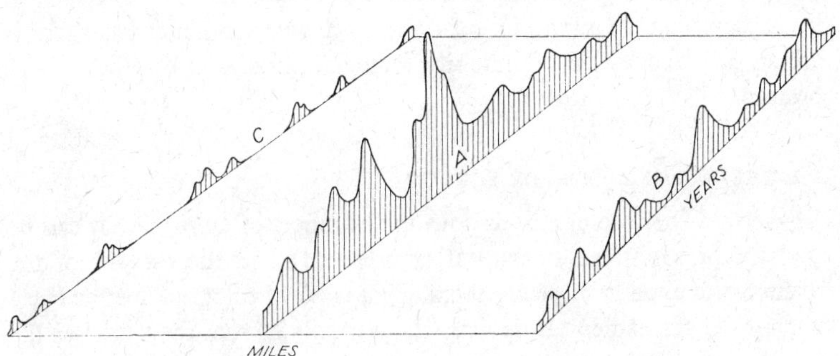

Fig. 1.03. Fluctuations in the abundance of the hypothetical animals that live at localities A, B, C in Figure 1.02. For further explanation see text. (After Andrewartha and Birch, 1954)

explained their distribution. It seems logical to define the idea of distribution by saying that the boundary of the distribution is that place where the abundance decreases to zero. This is why I say that distribution and abundance should not be treated as two separate ideas but merely as different aspects of the same idea.

With both *T. imaginis* and *A. cruciata* the systematic counting that we did formed the essential basis for the whole study. The problem of estimating the numbers in a natural population is basic for most field work in ecology. The emphasis may shift to different aspects: with *T. imaginis* the emphasis was chiefly on fluctuations in abundance from year to year; with *A. cruciata* this was important also but the emphasis was rather more on their distribution; on another occasion the emphasis might be on the relative abundance of a number of species and their relationships to each other, as in the studies of communities that Elton (1927) wrote about; or one might wish to compare the relative abundance of the animals from place to place in some small part of their distribution – whatever the aspect of ecology that is being emphasized, the first essential for good work is to know how many animals there are in the area that you are studying. That is why I have devoted the whole of Chapter III to the principles and techniques of estimating the numbers of animals in natural populations.

And that is why I think that the best definition of animal ecology is to say that *animal ecology is the scientific study of the distribution and abundance of animals*. At first sight this may seem to be a narrower definition than those that are commonly found in text-books of animal ecology. But this is not so as I shall show in sections 2.1 and 6.4. In order to explain the distribution and abundance of an animal you have to take into account all the important relationships between other animals which are emphasized by the community ecologists, and a lot of other things besides.

1.2 THE BROAD BASES FOR ECOLOGY

I have said that ecology deals with the numbers of animals that can be counted or estimated in natural populations; and the essence of the science is to explain these numbers. That is why ecologists need to learn at least the fundamentals of statistics, not only so that they can use the routine statistical methods to analyse the results of their experiments, but also in order that they may learn to think about their work in terms of probabilities. A knowledge of the behaviour of animals and their physiology is also a necessary part of the background for studying ecology. I hope that the ways that these skills are used in ecology will become clear in the course of this book. It is not so easy to write about the need for the arts of the naturalist – although this is perhaps the most important quality of all. It is best conveyed by the teacher, verbally and by precept. There is a passage from Ford (1945) which makes the point nicely.

> We have three butterflies which are limited by geological considerations, being inhabitants only of chalk downs or limestone hills in south and central England and they may reach the shore where such formations break in cliffs to the sea. These are the silver-spotted skipper, *Hesperia comma*, the Chalk-hill Blue, *Lysandra coridon* and the Adonis Blue, *L. bellargus*. . . . The two latter insects are further restricted by the distribution of their food plant, the Horse-shoe Vetch, *Hippocrepis comosa*, and possibly by a sufficiency of ants to guard them. Yet any of the three may be absent from a hillside which seems to possess all the qualifications which they need, even though they may occur elsewhere in the immediate neighbourhood. This more subtle type of preference is one which entomologists constantly encounter, and a de-

tailed analysis of it is much needed. A collector who is a careful observer is often able to examine a terrain and decide, intuitively as it were, whether a given butterfly will be found there, and that rare being, the really accomplished naturalist will nearly always be right. Of course he reaches his conclusions by a synthesis, subconscious as well as conscious, of the varied characteristics of the spot weighed up with great experience; but this is a work of art rather than of science, and we would gladly know the components which make such predictions possible.

1.3 HOW TO WRITE ABOUT ECOLOGY

It is an essential part of science that the scientist who has made a discovery should be able to tell his colleagues exactly what he did, how he did it and what are the conclusions to be drawn from his work. The clear, concise writing that is demanded of a scientist does not come naturally but only after many trials and much practice. The young scientist should avoid the mistake of copying the jargon that characterizes far too much of the 'scientific literature'. He would be wiser to study the masters of non-scientific literature because much the same rules govern the writing of simple, lucid English whether you are describing the animals in a wood or the actors in a play. Even the choice of words need not be as different as some scientists would make it. I have gathered together, in the next few pages, several pieces of sound advice from literary men as well as from scientists on this important subject.

Somerset Maugham (1948, p. 40), discussing the way that he set about attaining a style which would satisfy him, wrote:

> On taking thought it seemed to me that I must aim at lucidity, simplicity and euphony. I have put these three qualities in the order of importance I assigned to them. . . . You have only to go to the great philosophers to see that it is possible to express with lucidity the most subtle reflections. . . . There are two sorts of obscurity that you find in writers. One is due to negligence and the other to wilfulness. People often write obscurely because they have never taken the trouble to learn to write clearly. This sort of obscurity you find too often in modern philosophers, in men of science and even in literary critics. . . . Yet you will find in their works sentence after sentence that you must read twice in order to discover the sense. Often you can only guess at it, for the writers have evidently not said what they intended. . . . Some writers who do not think clearly are inclined to suppose that

their thoughts have a significance greater than at first sight appears. It is flattering to believe that they are too profound to be expressed so clearly that all who run may read, and very naturally it does not occur to such writers that the fault is with their own minds, which have not the faculty of precise reflection. Here again the magic of the written word obtains. It is very easy to persuade oneself that a phrase that one does not quite understand may mean a great deal more than one realizes. From this there is only a little way to go to fall into the habit of setting down one's impressions in all their original vagueness.

Leeper (1941, 1942) commented on two faults which, he said, often make the writings of scientists difficult to understand.

> . . . I have become steadily more convinced that the abstract noun is most to blame for the common heaviness. The first step on the downward path is to write 'solution takes place' instead of 'the substance dissolves'. The most important place is given to a noun instead of a verb, while the verb collapses to the emptiness of 'take place' or 'occur'.
>
> Once the first downward step is taken and the abstract noun is accepted as the pivot-word of the sentence, the second step is easy. This is the use of sets of two or more nouns, only the last of which has the full status as a noun, while the preceding nouns qualify it. The double noun, of course, is part of our language. Some double nouns, like *boatrace* and *newspaper*, have even been accepted as single words, while many others are commonly used without offence, like *vapour pressure* and *spot test*. But a writer of any sensibility will use them sparingly, and will avoid double nouns which are not in every day speech, such as the ugly *fertility decline* or *phosphate source*. Gowers (1948) has truly written, 'This unpleasant practice is spreading fast and is corrupting the English language.' He quotes Dunsany's satirical remark that we may use the frequency of multiple nouns in a given passage for dating our *language decay progress*. A treble noun like this is always inexcusable, yet treble or even quadruple nouns are common in our journals.
>
> The vice of this habit is not that experts think it is ugly, but simply that it makes reading difficult. We do not know at first whether we are reading about *language*, *decay* or *progress*, and we must waste some seconds in reading the sentence over in order to solve the puzzle.

Weiss (1950) mentioned a fault that may make the writings of a scientist not only difficult to comprehend but also misleading. There is a tendency, he said, to divest a phenomenon of its natural complexity by giving it a high-sounding name. Then the next step, which comes all too easily, is to slip into the habit of (or mislead the young into) thinking that the high-sounding name describes or explains the nature of the phenomenon (sec. 5.111).

Elton (1927, p. 6) protested against this fault in ecological writings.

> Various schemes have been proposed for the classification of animal communities, some very useful and others completely absurd. Since, however, no one adopts the latter, they merely serve as healthy examples of what to avoid, namely, the making of too many definitions and the inventing of a host of unnecessary technical terms. . . . After all what is to be said for the scientist who calls the community of animals living in ponds 'tiphic association', or refers to gardening as 'chronic hemerecology'?

Orwell's (1946) five rules are a useful guide for scientists. I repeat them below, not quite literally.

(1) Never use a long word where a short one will do.
(2) If it is possible to cut out a word always cut it out.
(3) Never use the passive where you can use the active.
(4) Use technical words sparingly and correctly; never use a word that cannot be understood by a scientist working in a related field.
(5) Break any of these rules rather than say anything barbarous.

1.4 FURTHER READING

There have been a number of books written about or around the subject of 'community ecology'. They include Elton (1927), Allee *et al.* (1949), Dice (1952), Clarke (1954), Odum (1953), Dowdeswell (1952) and MacFadyen (1957). The concepts of 'community', 'niche', 'habitat', 'ecosystem', 'pyramid of numbers' and 'community energetics' which are fundamental to the theory of 'community ecology' are discussed in these books. Elton (1927) is a classic and is still the best short account of the subject. This book foreshadowed most of the ideas that are now current. It can be commended to anyone who is interested in the development of modern ecology. The modern idea of 'habitat' is discussed in Elton (1949) and Elton and Miller (1954). There is a good discussion of the modern idea of 'community energetics' in Clarke (1954, Chap 13), and in Odum (1959) second edition.

The beginner would do well to follow up these suggestions for further reading in conjunction with his reading of this book.

CHAPTER II

Environment

2.0 INTRODUCTION

In section 1.1, I said that the scope of ecology was best defined as the scientific study of the distribution and abundance of animals, and that distribution was merely another aspect of abundance. In this chapter I want to examine the consequences of this idea and consider how to explain the numbers that occur in natural populations. Figures 2.01, 6.06 and 7.01 illustrate the numbers that were counted in three populations. The feature that they have in common is that the numbers fluctuated widely and erratically. This is typical of most of the natural populations that have been studied. It is clear that if I knew what caused the population to increase and decrease in this way, and could explain the magnitude of the fluctuations, I would have resolved the problem with which I began. I shall start with the simple idea that when births exceed deaths the numbers in the population increase, and when deaths exceed births the population must decrease. If the *rate of increase* is called r, the birth-rate b and the death-rate d we may write $r = b - d$; and it follows that r is positive when b is greater than d and negative when b is less than d. The birth-rate depends on the speed of development (from birth to maturity) and the fecundity of the individuals that make up the population; and the death-rate depends on the length of life of the individuals. Fecundity refers to the number of living progeny produced by a female. In egg-laying species eggs that have been laid are counted as progeny.

When the problem has been stated in this way it becomes amenable to analysis. Experiments may be done to show how temperature, moisture, food, 'crowding' etc., may influence the speed of development, fecundity and length of life of the animals (Chap. XI; Birch, 1953a, b, c). Then the population may be counted in the field and the results related to measurements of weather, food, crowding or whatever may seem to be important. It is convenient to have a general name that connotes everything that may influence the speed of development, fecundity and length of life of the animals that one studies. The name that has been chosen is 'environment'.

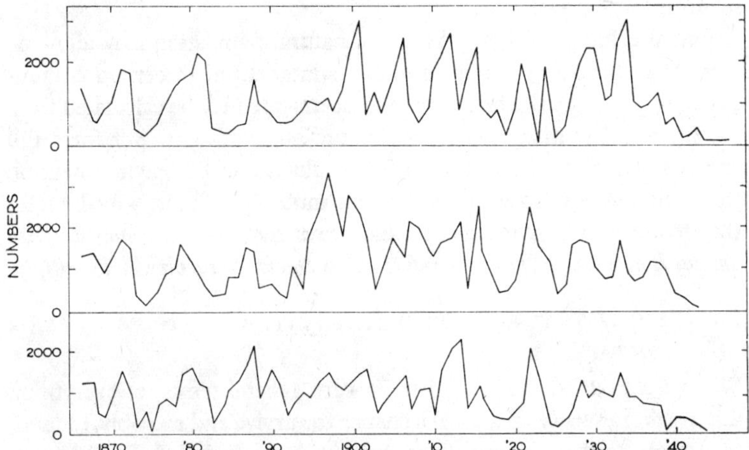

Fig. 2.01. Fluctuations in the numbers of red grouse shot on three moors in Scotland. (After Mackenzie, 1952)

When we do experiments on, for example, speed of development, we may, as in experiment 11.11, estimate the time that an animal takes to complete its life-cycle. We cannot rely on a single observation because animals vary; so we have to use a sufficiently large sample to represent the population adequately; we summarize the results by calculating the mean and its standard deviation. Similarly with experiments on fecundity the result is expressed as the number of eggs laid by the 'average' individual; and we specify just how representative our estimate of the 'average' is by calculating its standard deviation.

But we need to express these statistics differently in order to bring out their relationship to the 'abundance of the population' which is the idea that I started with. The way to do this is to think of the mean speed of development and mean fecundity as providing an estimate of the average individual's '*chance to multiply*'; and similarly the mean length of life gives a way of estimating the average individual's '*chance to survive*'. The average individual's 'chance to survive' is merely another way of speaking of its '*expectation of life*' which is a familiar idea to anyone who has contemplated taking out an insurance policy on his life.

The actual number of births in the population is got by multiplying the average individual's 'chance to multiply' by the number of individuals in the population; and similarly the actual number of survivors is got by multiplying the average individual's 'chance to survive' by the number of individuals in the population. The death-rate is the complement of the survival-rate.

Actual counts of the numbers in a natural population may allow no more than the rate of increase to be estimated, or, in certain circumstances (sec. 3.23), the birth-rate and death-rate may be calculated from the figures. For many purposes this may be sufficient. But for a full analysis it is necessary to think of the influence of the environment on the individual's chance to survive and multiply. This is why I prefer the definition of environment which says that *the environment of an animal is everything that may influence its chance to survive and multiply.*

2.1 THE IDEA THAT ENVIRONMENT IS DIVISIBLE INTO FOUR MAJOR COMPONENTS

When I set out to compile a list (in very general terms) of everything that might influence an animal's chance to survive and multiply I found, first of all, certain obvious items that had to be included. Food is one; so is weather because every animal wherever it is or whatever it may be doing must be exposed to some degree of temperature, moisture, light, etc. It is also obvious to anyone who has observed nature that scarcely any animal is free from the risk of being eaten or parasitized by some other kind of animal; and most animals also run at least some risk of infection from disease-causing plants and viruses; so all these must go in too. Then we have the example from human populations in the slums that when there are too many mouths to feed some have to go hungry. Occasionally we can observe the same thing with populations of other sorts of animals that outgrow their resources of food or something else that they need. The individuals in such populations have less chance to survive and multiply than they would if there were fewer of their own kind associated with them. The opposite probably happens more frequently, though it may be less obvious to the ordinary observer. The animals in sparse populations may have less chance of finding a mate or avoiding a predator or other hazards than they would if there were more of their own kind associated with them.

The last item that I added to the list is not so obvious as the ones that I have already mentioned. I arrived at it by thinking along the following lines: A beetle that spends the winter hibernating underground may have a better chance of surviving in an area where the soil is deep than in a place where there is a hard layer close to the surface that prevents the beetle from going deep enough to escape the frost. A rabbit living in a wood that also harbours a fox may live longer and produce more offspring if there is also a hollow log in the area. The advantage conferred on the rabbit by the presence of the hollow log depends not only on the presence of the fox. The advantage may be greater if there are not many other rabbits seeking shelter in the same log; also the advantage would be greater if the rabbit did not have to travel too far from shelter to seek food. Or, to take a third example, a deermouse may have inherited certain patterns of behaviour that enable it to live along the margins of a wood but not in the depths of the wood or in the adjacent meadow. An ecologist may be able to recognize the deermouse's requirements without being able to relate them to any other component of environment as he could with the deep soil required by the beetle or the hollow log required by the rabbit (see the passage quoted from Ford in sec. 1.2). Even so, he has to take this into account when he seeks to explain the distribution and abundance of the deermice. It is clear from what has been said in this paragraph that an animal's chance to survive and multiply may depend on the sort of place that it has to live in. If 'habitat' had not already been given a very different meaning for use in a very different context (sec. 1.0, Andrewartha and Birch, 1954, p. 28) this would have been a nice word to coin. But it is too late. Andrewartha and Birch (1954), being unable to think of a better name, called this component of environment 'a place in which to live'.

This list comprises four major categories which can be recognized by the way that they influence an animal's chance to survive and multiply. They are: (*a*) Weather, (*b*) Food, (*c*) Other animals – and pathogens, (*d*) A place in which to live. With the proviso that food for carnivores be placed in (*b*) along with all other sorts of food, and not in (*c*), the boundaries of these categories seem to be clearly marked and in no need of further definition.

I have found that I can think of the environment of any animal in terms of these four components and the *interactions* between them. By interactions I mean the sort of relationships that I described above when I mentioned that the log increased the rabbit's chance to survive and multiply more when (*a*) there was a fox in the area, (*b*) the log was close to a source of food for the rabbit, (*c*) there were not many other rabbits seeking shelter in the same log. The interactions in this instance are between 'a place in which to live' and (*a*) 'predators', (*b*) 'food' and (*c*) 'other animals of the same kind'. In subsequent chapters I shall mention many other examples of interactions between different components of environment. You have only to think about the multiplicity of the interactions that can occur to appreciate that this way of approaching the study of ecology need miss none of the relationships that are brought out by the direct study of communities. And it has the advantage of focusing attention on those that are essential (sec. 6.4).

With respect to the third component of environment, instead of saying that an *animal's chance to survive and multiply* depends on the number of *other animals of its own kind* that are associated with it, I might have said that the *birth-rate* and *death-rate* of the population depended on the *density of the population*. If I were using the concept of environment merely to describe what I had observed the second statement might be adequate; but if I want to *explain* my observations the first statement is better. I have given one reason for this above.

A second reason is that in order to explain a birth-rate or a death-rate directly in terms of environment we must be able to talk about the environment of a population. This raises the difficulty that the *density* of the population has to be regarded as part of the environment of the population. This seems to be a rather circular sort of argument, and in any case, it leads to difficulties which are best avoided by thinking of the environment of an individual, as I have done in my definition. This allows us to make a clear distinction between the *animal* and its *environment* which is helpful.

2.2 FURTHER READING

My definition of environment is different from that which is given in some other books. These other meanings may be studied in the references that I gave in the back of Chapter I.

In section 1.2 I said that the ecologist should know some statistics. In this chapter I have introduced the statistical terms 'mean' and 'standard deviation'. In subsequent chapters I shall have to use more of the ideas and terms of statistics. I have found that the 'Pelican' by Moroney, *Facts from Figures*, provides a good and relatively painless introduction to this subject for those who are not already familiar with it. Another good book is Mather, *Statistical Analysis in Biology*.

CHAPTER III

Density

3.0 INTRODUCTION

Ecologists commonly speak of the *density* of a population when they wish to refer to the number of animals in a particular area. Broadly, there are two ways of measuring the density of a population. If it is possible to count or estimate all the animals in the area it is proper to speak of the result as the *absolute density* of the population. But there are occasions when it is sufficient to know in what ratio one population exceeds another without knowing what is the actual size of either population. This is especially true when one is studying fluctuations with time in a population living in the same place. Sometimes technical difficulties may make it impossible to measure the absolute density of the population no matter how desirable this may be, and to estimate *relative densities* may be the best that one can do.

3.1 THE MEASUREMENT OF RELATIVE DENSITY

The characteristic feature of all methods for measuring *relative density* is that they depend on the collection of samples that represent a constant but unknown proportion of the total population. The sample may be collected by a conventional trap, such as a fly-trap or a mouse-trap, or by any other device that will serve the purpose. For example, when Browning (1959) needed to measure the fluctuations in a population of mealybugs that were living on the leaves of orange trees in an orchard in South Australia he collected samples of leaves and counted the mealybugs on them. It was impracticable to take the leaves strictly at random but Browning worked out a method that allowed him to take samples of several hundreds of leaves approximately at random except for the restriction that all the leaves were taken from the lower half of the tree. This method depended on the assumptions that (*a*) the leaves were picked at random, (*b*) the total number of leaves on the tree did not vary, (*c*) there was no change in the relative proportions of the mealybugs on the lower and upper parts of the tree. Provided that these three conditions are fulfilled the mean number of mealybugs on a leaf represent a constant proportion of the population on the tree. The actual density of the population could have been estimated if the number of leaves on the tree had been known. But it was impracticable to count them. It was also unnecessary because in this study relative densities were sufficient.

The *suction trap* used by Johnson (1950a) to catch aphids and other small insects floating in the air consisted essentially of a large vertical funnel, a fan driven by an electric motor, and a jar in which the insects were collected. The fan was designed to suck a constant volume of air through the funnel, about 19,000 cubic feet an hour. Such a trap, running for a known time, would collect all the aphids from a known volume of air. With sufficient number of traps appropriately distributed in the air it becomes merely a matter of simple arithmetic to estimate the actual number of aphids in the air over a particular area at a particular time. The rest of the population are on plants. If the number in the air represented a known proportion of the ones on the plant then the calculation could be extended to give an estimate of the absolute density of the population. If, on the other hand, the numbers in the air represented a constant but unknown proportion of the total population the method could provide estimates of relative density. Johnson (1952a, b, 1955) has used the records more for the study of dispersal than of density (sec. 4.11). Formerly it had been thought that the proportion of the population in the air depended on the immediate or very recent weather stimulating the aphids to fly. Although this still seems to be important Johnson has shown that the numbers in the air depend much more than had been thought on the actual density of the population on the plants.

Estimating changes in the numbers of aphids on a crop of, say, potatoes or beans may be a matter of considerable economic importance but it presents a tricky problem because the numbers may be large and they may be distributed unevenly. Banks (1954) used the following method in a crop of beans. The beans were planted in rows. A number of stems were chosen at random in each row and 'rated' into five categories merely by looking at them and judging the extent of the infestation. The five categories were zero, very few, few, many and very many. Ten stems from each class were cut off at the base and the aphids on them were counted. From the average number of aphids on a stem in each category and from the frequency of stems of each category in the field a weighted mean number of aphids per stem was estimated. In a series of 9 weekly estimates made in this way the mean number of aphids per stem increased from 20 at the beginning to 14,000 towards the end of the period. The number of stems per acre was not known, so these figures cannot be converted into an estimate of absolute density.

The light-trap that Williams (1940) used consisted essentially of an incandescent electric globe, a funnel and a jar. The light attracted night-flying Lepidoptera and other insects; the funnel directed them into the jar where they were killed. Williams ran the trap for a number of years at Rothamsted. The numbers of insects in the trap varied greatly from night to night. A substantial part of this variation was related to fluctuations in certain components of the weather which probably influenced the moths' activity in seeking out the trap. This result illustrates the general rule that the numbers caught in any trap depend partly on the density of the population that is being sampled and partly on the activity of the animals in seeking out the trap. It is desirable to find some way of distinguishing between these two components before using the results as an estimate of relative density. (See sec. 5.15 for an example of how this was done with *T. imaginis*.)

3.2 THE MEASUREMENT OF ABSOLUTE DENSITY

3.21 *Counting the whole population*

The most direct way to find out how many animals are living in an area is to count them all. Errington (1945) kept a complete record during 15 winters of the number of bobwhite quail, *Colinus virginianus*, on 4,500 acres of farmland in Wisconsin by counting either the birds or their 'sign'. The sign left by the birds in the snow supplied a lot of information apart from their mere numbers. Similarly with the muskrat, *Ondatra zibethicus*, Errington (1943, 1946) was able to count all those living in certain marshes in Iowa. Again the counting was done largely by reading sign.

Kluijver (1951) reported a 40-year study of the ecology of the great tit, *Parus major*, in 130 acres of mixed woodland in Holland. The tits normally use holes in trees for nesting during summer and roosting during winter. But they will readily use artificial nesting boxes instead. This behaviour allowed Kluijver to inspect the birds at night, to mark them with leg-bands and to count them all.

The Australian magpie, *Gymnorhina dorsalis*, is unusual among birds in that 'tribes', consisting of from 2 to 10 birds, live permanently in the same territory, defending it against all comers at all times of the year (sec. 6.122). Carrick (1959) made a trap by dividing a wire cage of about one cubic metre into 9 compartments. The central compartment housed a bird from foreign territory. The peripheral cages were fitted with trap-doors. The foreigner attracted the local birds which entered the cages and were trapped. The local birds were marked with coloured leg-bands and released. Each year the nestlings were marked before they left the nest. In this way Carrick was able to keep a complete and detailed record of the whole population.

Usually there is a better chance of counting the whole population when dealing with vertebrates, especially birds and rodents, than with insects and other invertebrates. This method has many advantages but the occasions on which it can be used are relatively infrequent even with vertebrates. Usually one has to rely on some method of sampling in which only a small proportion of the population is seen. Broadly there are two methods available. Either one counts every animal in a quadrat that represents a known *proportion* of the total population. Or one captures, marks and releases a known *number* of animals. The total population is estimated from the proportion of marked ones that are subsequently recaptured. Of course, the details vary widely according to the behaviour of the particular animals that are being studied. I shall discuss these two methods in sections 3.22 and 3.23.

3.22 *The use of quadrats*

In the simplest form of this method the quadrats are merely small areas chosen at random from a larger area which contains the whole population. The conditions that must be fulfilled are that the population of the quadrat must be known exactly and the area of the quadrat relative to the whole area must be known. Details of the devices that have been used for different species vary greatly. I shall mention only a few.

Salt *et al.* (1948) estimated the number of arthropods in an acre of soil by cutting out cylinders of soil 4 inches in diameter and 12 inches deep. They invented a special technique for pulverizing the soil without destroying the animals in it. The soil was then washed through a series of sieves and the fractions containing plant debris were further sorted by a flotation process using an interface between paraffin and an aqueous solution of magnesium sulphate. On one occasion Salt *et al.* found that a sample of 20 quadrats (i.e. 20 cylinders of soil 4 inches in diameter) contained on the average 2,138 animals per quadrat. Multiplying this figure by 498,960 (the number of quadrats that could be taken from one acre) gives an estimated population of 1066·6 million arthropods per acre. Of these 666 million were mites and 379 million were insects.

Often it is possible to make use of some peculiarity in the behaviour of the animal in order to collect them from or count them in the quadrats. Madge (1954) used quadrats one foot square to estimate the density of the population of *Oncopera fasciculata*. The caterpillars live in vertical burrows in the soil under pastures of clover and grass. The mouth of the burrow is difficult to find because it is well camouflaged. Even clearing away the herbage does not help because this blocks the mouth of the burrow and makes it invisible. But if the herbage is cleared away and then a flat board is pressed down on the soil the caterpillars will clean out their burrows and spin a silken cap to them. When the board is lifted next day the mouths of the burrows stand clearly revealed and can be counted easily.

The small fly, *Spaniotoma minima*, breeds in sewage filter-beds near Leeds. The eggs, larvae and pupae are found in the filter-beds but the adults emerge into the air to mate. Mating does not occur casually on or near the surface of the filter-bed but only after an elaborate nuptial flight that involves the formation of a nuptial swarm. The flies always seek to escape from the surface of the filter-bed. Consequently Lloyd (1941) was able to trap them by placing a funnel-like frame one foot square on the surface of the filter-bed which led the flies upward into a jar in which they were trapped. Multiplying the number of flies caught in a trap by the surface area of the filter-bed in square feet would give an estimate of the absolute density of the population.

The sheep tick, *Ixodes ricinus*, takes about 3 years to complete its life-cycle. Most of this time is spent near the surface of the ground usually concealed in a dense mat of vegetation and decaying organic matter where it is difficult to find. But during the spring each year there is a period of several weeks during which the ticks come up for perhaps a day or two at a time to sit on the tips of grass-stems and other exposed places. If a sheep brushes it at this time the tick will cling to the sheep. Milne (1943) found that the ticks would also cling to a woollen blanket if it were dragged over the grass. Of course not all the ticks were exposed at once but by 'dragging' the same area repeatedly Milne considered that he could collect all or nearly all the ticks. He chose a number of quadrats of known area and 'dragged' them thoroughly. In this way he estimated the total population of ticks on a particular farm at between 140,000 and 230,000.

The reliability of the estimates made from counting the numbers in quadrats depends on how representative the quadrats are of the whole population and on how close one gets to seeing all the animals in the quadrat. The latter question is a matter of technique and must be resolved on its merits for each particular species. Salt *et al.* pointed out that their method was likely to miss any animal that could pass through a 0·1-mm. sieve; they thought that the number missed in this way might be quite substantial. Milne could count only the ticks that had emerged from the 'mat' and were picked up by his blanket. He 'dragged' the same area repeatedly during the active season and probably left very few behind. Madge could count only the *Oncopera* that repaired their tunnels when he lifted the board. But he could check his results by digging out the soil and sieving it. It seemed that he would miss very few provided that he sampled when the larvae were not likely to be moulting.

The question of how well the sample of quadrats may represent the population is largely a statistical matter and may best be answered in section 3.33 after I have discussed the 'patchiness' with which most sorts of animals tend to be distributed in even quite small and seemingly uniform areas. The less evenly the population is distributed in the area the more quadrats will be required to represent it adequately. If the 'patchiness' of the population follows a particular pattern it may be necessary to recognize this in placing the quadrats. The statisticians have developed a special technique which is called 'stratified sampling' to meet these requirements. But this is outside the scope of this book.

3.23 *The method of capture, marking, release and recapture*

If, from an area that has been selected for study, M individuals have been captured, marked and then released again, and if on a second occasion N individuals are captured, including R marked ones, then the population in the area may be estimated as:

$$P = N \times \frac{M}{R} \tag{3.1}$$

The idea expressed by this equation is simple enough in principle but it implies two assumptions that need to be discussed.

(1) Equation 3.1 implies that the marked ones after they were released distributed themselves homogeneously with respect to the unmarked ones which had not been caught. The equation also implies that on subsequent occasions any marked one has the same chance of being recaptured as any unmarked one has of being captured.

(2) Equation 3.1 also implies that the recapturing is done immediately after the releasing, or at least before there has been time for any marked one to die or leave the area, or for any immigrant to enter the area.

In practice this cannot be done because the marked ones must be allowed time to mix with the unmarked ones.

There are also two technical difficulties that should be mentioned.

(1) When traps are used, especially in an area that is not limited by sharp ecological boundaries, difficulties arise with respect to the placing of traps relative to the boundaries of the area.

(2) It is difficult to attribute a precise variance to the estimates that are given by this method because the calculations are tedious and difficult.

Dowdeswell *et al.* (1940) estimated the density of a population of the butterfly *Polyommatus*. The butterflies were caught with a net, marked and released in the same place that they were caught. Jackson's (1939) method of capturing and releasing the tse-tse fly, *Glossina morsitans*, was similar. There would seem to be little risk of disturbing the distribution of the marked relative to the unmarked when this method is used. But often it may be necessary to use some sort of trap. If the trap lures the animals from any distance it results in an artificial aggregation of a part of the population. These individuals are marked and the problem is to get them back into the population so that they are distributed in the same way as the unmarked ones. It is not sufficient merely to scatter them at random to the points of the compass or to any other geometrical co-ordinates because the unmarked part of the population may not be distributed in this way. All that can be done is to release them and hope that, in settling down, they will conform to the same pattern of distribution as the unmarked ones.

Evans (1949) exposed a number of traps in a building where there was a population of *Mus musculus*. Certain individuals were captured many times whereas others were not caught once. The differences were too great to be attributed to chance. It was clear that certain mice were much more likely to be caught than others. This placed an intolerable bias on the estimates of the density of the population and the exercise had to be abandoned. Even when more direct methods are used for capturing the sample there may be a risk of bias from this cause. For example, if the more sluggish animals are more easily caught than the more lively ones, then the more sluggish would also be more easily recaptured and this would tend to place a bias on the estimates of density. Even if it were true only that the older ones were more easily caught than the young ones, this would introduce bias because the marked ones must be older than the unmarked. There is no standard way to get over this difficulty; each case must be treated on its merits even to abandoning the exercise if the problem cannot be solved.

If the marked redistributed themselves quite homogeneously with respect to the unmarked the recapturing could be done systematically or haphazardly over the whole area or part of it without any risk of introducing bias to the results. But in so far as the marked are not distributed homogeneously with respect to the unmarked (and in practice they rarely will be) the only way to reduce the risk of bias is to randomize the positions of the recaptures over the whole area.

Equation 3.1 implies that the recapturing is done immediately after releasing but in practice some delay must be allowed while the marked animals are given time to disperse and settle down. Consequently it is necessary to extrapolate backwards from the date when they are recaptured to the date when they were released. This makes it necessary to recapture on a number of occasions instead of only once, or alternatively to mark on a number of occasions and recapture once. The former was called the 'positive' and the latter the 'negative' method by Jackson (1939); both lead to the same result within the limits of sampling errors. The calculations are best explained by reference to a particular example and I shall use some figures that were published by Jackson (1939) for a population of tse-tse fly, *Glossina morsitans*.

The figures are set out in Table 3.01. First consider the lower part of the table covering the period from t_0 to t_6. These are the data required for the 'positive' method of estimating the number in the population on date t_0. The table shows that on date t_0 1,558 flies were captured, marked and released. On date t_1 a total of 1,547 flies were caught, including 101 that had been marked and released on date t_0. On date t_2 the catch totalled 1,307, including 81 of the marked ones. The numbers in these two columns vary from day to day. In order to simplify the calculations we correct the figures for recaptures to what they would have been if 100 had been marked on date t_0 and 100 captured on each successive date. We call these corrected figures $y_1, y_2 \ldots y_6$. They are calculated as follows:

$$y_1 = \frac{101}{1} \times \frac{100}{1{,}558} \times \frac{100}{1{,}547} \qquad (3.2)$$

$$= 0{\cdot}4190$$

We require to find by extrapolation from this series y_0 which we think of as the (corrected) number that would have been recaptured on t_0 if we had been able to recapture immediately after releasing. In order to recognize the difference between this theoretical estimate and the empirical series of y we call our estimate a_0 instead of y_0. In order to

TABLE 3.01

Jackson's (1939) data for *Glossina morsitans* (slightly adjusted)*

Date of capture or recapture	*Marked and released*	*Total captured*	*Recaptured*	*Number of flies corrected recaptures* y	*Estimated population*	
					Negative	*Positive*
t_{-6}	1,262		1	0·0051		
t_{-5}	1,086		5	0·0296		
t_{-4}	1,299		17	0·0840		
t_{-3}	1,401		28	0·1283		
t_{-2}	1,198		48	0·2569		
t_{-1}	1,183		70	0·3798		
t_0	1,558	1,558			12,800	13,200
t_1		1,547	101	0·4190		
t_2		1,307	81	0·3978		
t_3		603	24	0·2555		
t_4		1,081	31	0·1841		
t_5		1,261	10	0·0509		
t_6		1,798	11	0·0393		

* Jackson's original figures show 1,582 marked on date t_0 for the positive method and 1,558 caught on date t_0 for the negative method. The discrepancy is due to 24 young unfed flies which, according to the conditions that Jackson had made, were eligible for marking but not for counting in the 'negative' catch. In order to simplify the arithmetic I have ignored these 24 young flies and adjusted the recaptures for the positive method as if 1,558 had been marked and released on day t_0.

estimate a_0 we first require to calculate r_+ which is the weighted ratio of $\frac{y_n}{y_{n-1}}$. When there are no more than 3 or 4 values for y (i.e. when recaptures were made on no more than 3 or 4 dates) it may be necessary to estimate r_+ from equation 3.3:

$$r_+ = \frac{y_2 + y_3 + \ldots + y_n}{y_1 + y_2 + \ldots + y_{n-1}} \tag{3.3}$$

But when there are sufficient values of y it is better to use equation 3.4 which carries the smoothing further:

$$r_+ = \sqrt{\frac{y_3 + y_4 + y_5 + \ldots + y_n}{y_1 + y_2 + y_3 + \ldots + y_{n-2}}} \tag{3.4}$$

In the present example the value for r_+ by equation 3.4 is:

$$r_+ = \sqrt{\frac{0{\cdot}2555 + 0{\cdot}1841 + 0{\cdot}0509 + 0{\cdot}0393}{0{\cdot}4190 + 0{\cdot}3978 + 0{\cdot}2555 + 0{\cdot}1841}}$$
$$= 0{\cdot}6494$$

The value for a_0 is given by equation 3.5:

$$a_0 = \frac{y_1 + y_2 + \ldots + y_{n-1}}{r_+} - (y_1 + y_2 + \ldots + y_{n-2}) \quad (3.5)$$

Because $\frac{y_1}{r_+}, \frac{y_2}{r_+}$, etc. may be regarded as estimates of y_0, y_1, etc., taking into account the weighted estimate of r_+, it is convenient to regard equation 3.5 as being equivalent to:

$$y_0 + (y_1 + y_2 + \ldots + y_{n-2}) - (y_1 + y_2 + \ldots + y_{n-2}) \quad (3.6)$$

The two terms in brackets cancel out. For the present example:

$$a_0 = \frac{0{\cdot}4190 + 0{\cdot}3978 + 0{\cdot}2555 + 0{\cdot}1841 + 0{\cdot}0509}{0{\cdot}6494} - (0{\cdot}4190 + 0{\cdot}3978 + 0{\cdot}2555 + 0{\cdot}1841)$$
$$= 0{\cdot}757$$

Now all the values of y, and therefore of a_0 have been corrected as if 100 flies were marked on date t_0 and 100 were captured on all subsequent dates. Therefore, substituting these values in equation 3.1, we have:

$$P \text{ (on date } t_0) = \frac{100}{1} \times \frac{100}{a_0} \quad (3.7)$$

For the present example:

$$P_0 = \frac{100}{1} \times \frac{100}{0{\cdot}757}$$
$$= 13{,}200$$

This is the estimate, by the positive method, of the population on date t_0.

Now consider the top half of Table 3.01 which sets out the data for the negative method. In this method we consider recaptures on date t_0 of flies that were marked and released on previous dates $t_{-1}, t_{-2}, \ldots t_{-n}$. Thus in column 4 the number 28 opposite t_{-3} indicates that of the 1,401 flies that were marked and released on date t_{-3}, 28 were recaptured on date t_0. Similarly 17 out of the 1,299 that were marked and released on t_{-4} were recaptured on t_0. The total catch on date t_0 including all those that had been marked on the six previous occasions was 1,558.

The equations for calculating $y_1, y_2 \ldots y_n$ are the same as in the positive method. For example:

$$y_{-4} = \frac{17}{1} \times \frac{100}{1{,}299} \times \frac{100}{1{,}558}$$
$$= 0{\cdot}0840$$

And r_- and a_0 are calculated from the same equations as in the positive method. For the present example:

$$r_- = 0{\cdot}5394$$
$$a_0 = 0{\cdot}780$$
$$P_0 = 12{,}800$$

In this method we think of a_0 as the (corrected) number that would have been recaptured on t_0 if they had been marked and released on that day and recaptured immediately before any of the marked ones had been lost by death or emigration. The two estimates of P_0 agree quite well.

Dowdeswell *et al.* (1940) invented a special trellis diagram for setting out the results obtained by the method of capture, mark, release and recapture. By merely inspecting the diagram it is possible to detect trends in the population due to births and immigration and deaths and emigration. The trellis also sets out, in easily accessible form, all the data required to calculate the actual population on any sampling date. It is not possible to complete the trellis diagram for the above example because we do not have sufficient information; the details in Table 3.01 are sufficient for only two columns of the trellis. The figures published by Dowdeswell *et al.* for *Polyommatus icarus* are rather more complete (Table 3.02).

In the body of the table the figures in roman type give the actual recaptures (equivalent to column 4 of Table 3.01); and the figures in italic type give the corrected values assuming 100 marked and 100 caught, as in column 5 of Table 3.01. A column starting at any date and running downwards to the left sets out the details required for the negative method. Each column is equivalent to the top half of Table 3.01 because the figures give the number recaptured on this date of those that were marked on some earlier date which is indicated at the top of the diagonal column. Thus about the middle of the table there is the figure 16 which indicates that on September 2, 16 butterflies were recaptured out of the 50 that had been marked on August 30. A column starting at any date and running downwards to the right gives the details required for the positive method. Each column is equivalent to the bottom half of Table 3.01 because the figures refer to the numbers that

TABLE 3.02

Trellis diagram showing changes in the population of *Polyommatus icarus*

	AUGUST 26th.	27th.	28th.	29th.	30th.	31st.	SEPTEMBER 1st.	2nd.	3rd.	4th.	5th.	6th.	7th.	8th.
Estimated number in population	344	234	•	148	169	241	•	160	150	•	•	68	35	•
Totals captured	40	43	•	13	52	56	•	52	50	•	•	15	20	7
Totals released	40	40	•	12	50	51	•	52	50	•	•	15	20	•

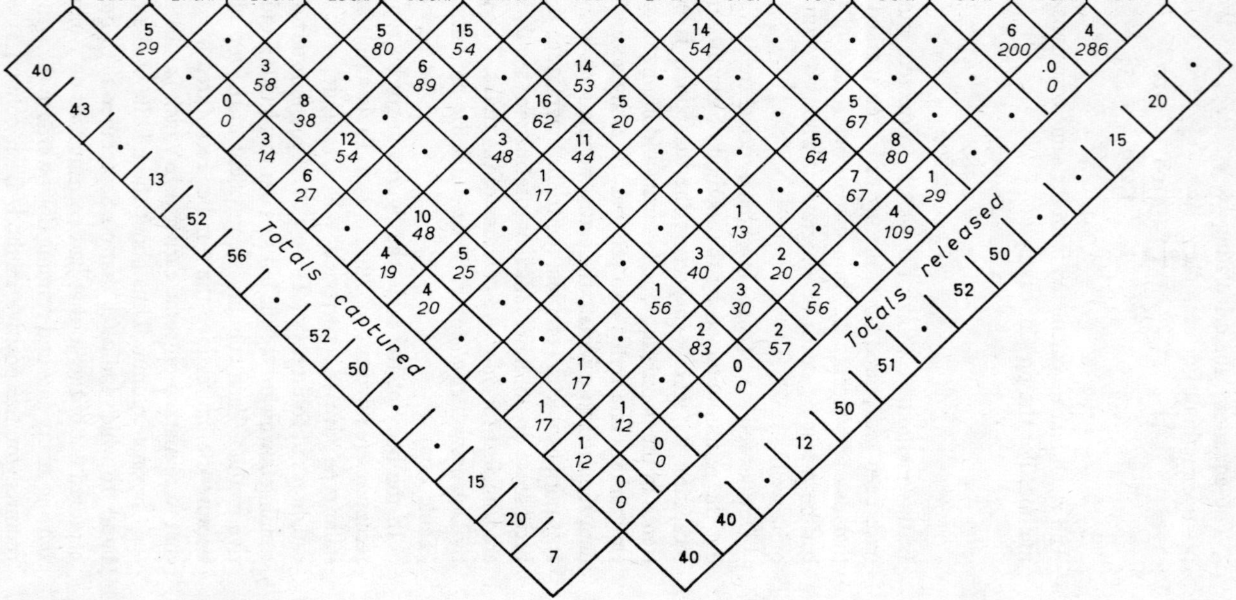

	26th.	27th.	28th.	29th.	30th.	31st.	1st.	2nd.	3rd.	4th.	5th.	6th.	7th.
27th.	5 / *29*												
28th.	•	•											
29th.	0 / *0*	3 / *58*	•										
30th.	3 / *14*	8 / *38*	•	5 / *80*									
31st.	6 / *27*	12 / *54*	•	6 / *89*	15 / *54*								
1st.	•	•	•	•	•	•							
2nd.	4 / *19*	10 / *48*	•	3 / *48*	16 / *62*	14 / *53*	•						
3rd.	4 / *20*	5 / *25*	•	1 / *17*	11 / *44*	5 / *20*	•	14 / *54*					
4th.	•	•	•	•	•	•	•	•	•				
5th.	•	•	•	•	•	•	•	•	•	•			
6th.	1 / *17*	1 / *17*	•	1 / *56*	3 / *40*	1 / *13*	•	5 / *64*	5 / *67*	•	•		
7th.	1 / *12*	1 / *12*	•	2 / *83*	3 / *30*	2 / *20*	•	7 / *67*	8 / *80*	•	•	6 / *200*	
8th.	0 / *0*	0 / *0*	•	0 / *0*	2 / *57*	2 / *56*	•	4 / *109*	1 / *29*	•	•	0 / *0*	4 / *286*

were recaptured on subsequent dates of those that were marked on this date. Thus of the 52 that were marked on September 2, 30 were subsequently recaptured, 14 on September 3, 5 on September 6, 7 on September 7 and 4 on September 8. The marginal totals in the row running downwards from the top left corner are equivalent to column 3 of Table 3.01 and the marginal totals in the row running downwards from the top right corner are equivalent to column 2 of Table 3.01.

Dowdeswell *et al.* were not able to count *Polyommatus* at regular intervals because they could not catch them during bad weather. Consequently we cannot use Jackson's methods for calculating a by extrapolating back from the calculated values of y because this method requires recapturing to be done regularly. Dowdeswell *et al.* estimated the approximate density of the population on successive sampling dates by a more direct use of equation 3.1. Their figures are given along the top of Table 3.02. The student should read the original paper to find out the details of their calculations.

When the recaptures have been made at regular intervals for a sufficient period the complete trellis diagram gives a complete picture of the trends in the population (exp. 10.25); but it is not necessary to have complete records to discover these trends. Consider Table 3.02. There is very little trend to be seen in the figures running down towards the right: apparently there were few new individuals added to the population during the whole period except perhaps about August 31. The figures in the columns running down to the left show a pronounced tendency to decrease. Apparently there was a steady and continuous loss by deaths and emigration during the whole period. These trends in the gains and losses are reflected in the size of the population which began with 344 members, declined steadily except for a small increase about August 31, and ended with 35.

Jackson (1939) showed how to separate births from immigration and deaths from emigration.

Let $D + E$ stand for the proportion of the population lost by death and emigration. Let $B + I$ stand for the proportion of the population gained by birth and immigration.

It might seem more usual to multiply $D + E$ and $B + I$ by 100 and think of them as rates *per cent*, but in the present context it simplifies the calculations if we leave them as they are and think of $D + E$ as a rate of loss per individual and $B + I$ as a rate of gain per individual. We use 'survival' to mean the proportion of the animals still living in the area where the marked ones were released.

We can write that the survival-rate per individual from date t_0 to t_1, is equal to:

$$\frac{\text{Number recaptured on date } t_1}{\text{Number marked on date } t_0} \times \frac{\text{Population on date } t_1}{\text{Number caught on date } t_1}$$

Expressing this in the usual terms of corrected recaptures (i.e. corrected as if 100 had been marked and 100 captured) we get:

$$\text{Survival rate } (t_0 \text{ to } t_1) = \frac{y_1}{100} \times \frac{P_1}{100} \tag{3.8}$$

Because the relationship $P \times a = 10{,}000$ (equation 3.7) is generally true it is also true for date t_1. Therefore equation 3.8 may be written:

$$\text{Survival rate } (t_0 \text{ to } t_1) = \frac{y_1}{a_1} \tag{3.9}$$

Now, in the negative method, y_1 relates to individuals marked and released on t_1 and actually recaptured on t_2; whereas a_1 relates to individuals marked and released on t_1 and assumed to be immediately recaptured on t_1. Therefore, for a smooth curve, the ratio of y_1 to a_1 is equivalent to r_-.

Because r_- is the proportion surviving from t_0 to t_1 the number that survives during this period is $P_0 r_-$ and the number lost by death and emigration is therefore $P_0 - P_0 r_-$. To express this as a rate per individual we divide by P_0. Thus the rate of loss per individual at date t_0 is given by:

$$D + E = 1 - r_- \tag{3.10}$$

To arrive at a similar expression for the rate of gain by births and immigration we reason as follows. The number that die or emigrate between t_0 and t_1 is given by:

$$P_0 (1 - r_-)$$

Therefore the number that remain on t_1 is given by $P_0 - P_0 (1 - r_-)$. Therefore the number that enter or are born in the area between t_0 and t_1 is:

$$\begin{aligned} & P_1 - [P_0 - P_0 (1 - r_-)] \\ &= P_1 - P_0 r_- \end{aligned}$$

This can be expressed as a rate per individual by dividing by P_1. Thus the rate of gain per individual at date t_1 is given by:

$$B + I = 1 - \frac{P_0 r_-}{P_1} \tag{3.11}$$

$$= 1 - \frac{a_1}{a_0} r_-$$

I have shown above (equation 3.9) that for a smooth curve:

$$r_- = \frac{y_1}{a_1}$$

Therefore $$B + I = 1 - \frac{y_1}{a_0}$$

Now, in the positive method, y_1 relates to individuals that were marked on t_0 and recaptured on t_1; whereas a_0 relates to individuals that were marked on t_0 and assumed to be recaptured immediately on t_0. Therefore the ratio of y_1 to a_0 is equivalent to r_+.

Therefore $$B + I = 1 - r_+ \tag{3.12}$$

If it is desired to estimate the rates of births, deaths, immigration and emigration separately then it is necessary to proceed as follows. First the area must be laid out as a large square that is subdivided into four smaller squares. Then, provided that the dispersal of the animals is not too large relative to the size of the small squares, we should be able to make the following assumptions: (*a*) The death-rate and birth-rate should be the same, subject to errors of random sampling, whether it is estimated from the large square or from a smaller one. (*b*) The rates of immigration and emigration to and from the small squares should be twice that to and from the large square because each small square has twice as much perimeter relative to its area as the large square.

If we mark, release and recapture over the whole area but keep the records for the small squares separately, we may write r_+, r_- for the large square and r'_+, r'_- for the means of the four small squares.

Then:

$$D + E = 1 - r_-$$
$$D + 2E = 1 - r'_-$$

therefore $$D = 1 - 2r_- + r'_- \tag{3.13}$$

and $$E = r_- - r'_- \tag{3.14}$$

By analogous reasoning it can be shown that:

$$B = 1 - 2r_+ + r'_+ \tag{3.15}$$

and $$I = r_+ - r'_+ \tag{3.16}$$

Because D has been defined as the death-rate per individual we may also think of D as the chance that an individual that is alive on date t_0 will be dead by date t_1. Therefore $(1-D)$ may be thought of as an expectation of life. It gives the chance that an individual that is alive on date t_0 will still be alive on date t_1.

If we could assume that D remains constant we could use $(1-D)$ to calculate the average duration of life of an individual. But this seems to be rather an unrealistic assumption because death-rate in the population is likely to change widely with the changing age-structure of the population. But the above methods can be used with the trellis diagram to follow changes in the birth-rate and death-rate in the population. In practice the assumption (*b*) above is not likely to hold very closely unless we happen to choose distances such that the side of a small square is equal to the average distance travelled by an animal during the experiment.

Boguslavsky (1956) described a method of capture, marking, release and recapture that is especially suitable for small populations. One individual is captured, marked distinctively and released. Subsequently samples of one are drawn repeatedly from the population and released again. On the first occasion that any individual is seen it is marked distinctively before being released. Suppose that in one area, after repeating this procedure a number of times, only 3 animals have been seen and marked. If the procedure is continued until 13 recaptures have failed to reveal any other animal than these three then the odds against there being more than 3 animals in the area are 9 to 1, always provided that all the animals in the area are equally likely to be captured. Similarly if 15 animals were seen and marked it would be necessary to repeat the sampling for 78 consecutive times without seeing a stranger in order to establish odds of 9 to 1 against there being more than 15 animals in the area. Although the arithmetic becomes rather cumbersome for larger numbers Boguslavsky showed how to calculate the number of negative trials that are necessary in order to establish any size of population at any level of probability. Subject to the restrictions mentioned above, this method seems promising for small local populations especially in wildlife studies. Still other methods have been developed for estimating the number of fish in a lake or other enclosed area. For an example of how these methods may be used see Gerking (1952).

3.3 THE MEASUREMENT OF 'AGGREGATION' IN NATURAL POPULATIONS

In his essay on 'population interspersion' Elton (1949) pointed out that no habitat is homogeneous. Particular sorts of animals may be restricted to particular 'minor habitats' scattered through the larger area. Thus in a mixed deciduous wood there may be a particular sort of insect that can live only on oak trees. This immediately imposes a 'patchiness' on the distribution[1] of the animals: they are to be found only where there is an oak tree. On close examination, any habitat, even such a seemingly uniform one as a grove of citrus, will reveal more or less heterogeneity with respect to the sorts of places where animals may live. The trees may differ with respect to genetical constitution, health, age, etc.; even those that are otherwise uniform must have a north and a south face. This heterogeneity in habitats is likely to be reflected in 'patchiness' in the way that the animals are distributed in the area.

But if one narrows the study-area down to a small and uniform area such as Elton called a 'micro-habitat' – the leaves on the western face of a tree, or 6 square inches under its bark – one will still find, more often than not, that the animals are distributed according to some pattern that is different from what would be expected if they had been scattered at random over the area. Often the explanation of the pattern lies in the behaviour of the animals or in some events in the history of the population.

I shall discuss several methods for measuring these 'patterns of distribution' in sections 3.31 and 3.32.

3.31 *A test for randomness: the Poisson series*

Imagine a uniform area subdivided into several hundred quadrats. Imagine that animals have been scattered over the area at random to each other and to the area. This means that each animal, as it was released, was just as likely to settle in any one quadrat as in any other and that its chance of settling in a quadrat was independent of whether this quadrat was already occupied by one or more animals or not. The frequency distribution setting out the number of quadrats with 0, 1, 2, 3, . . . animals on them is likely to conform, within the limits of random sampling, to the expansion of the binomial $(p + q)^n$ where p is the chance that an animal will settle in any quadrat; q is the chance that it will not; because the animal must either settle or not settle in that quadrat $p + q = 1$; n is the total number of animals. Provided that

[1] In this section 'distribution' is used with its statistical meaning. The relative frequency of quadrats containing 0, 1, 2, . . . animals forms a 'frequency distribution'. The context is quite different from that in Chapter I where 'distribution' was used in its ecological meaning to connote 'area'.

n is large, that p is small relative to q and that p is of the same order of magnitude as $\frac{1}{n}$ the expansion of the binomial $(p + q)^n$ approximates to the Poisson series which is given by:

$$P_x = e^{-m}\left(1, m, \frac{m^2}{2!}, + \dots \frac{m^x}{x!}\right)$$

In this expression P_x is the probability of getting 0, 1, 2, . . . animals in a quadrat. The frequency with which these numbers would be expected in an experiment in which N quadrats had been counted is given by multiplying the probabilities by N.

It is characteristic of the Poisson series that the variance equals the mean so that the ratio $\frac{S(x - \bar{x})^2}{(n - 1)\bar{x}} = 1$; and for a Poisson series the expression $\frac{S(x - \bar{x})^2}{\bar{x}}$ gives a good approximation to χ^2 with $n - 1$ degrees of freedom. Values of less than unity for the first expression indicate that the observed distribution from which it was calculated is less uniform, and values greater than unity indicate that the distribution is more variable than the Poisson series. Departures from unity in either direction increase the significance of χ^2. Methods for calculating the Poisson series and the tests for agreement with the observed series, as well as the meaning to be attached to the tests for significance, are discussed in section 10.1.

Table 3.03 shows how this method was used to measure the pattern of distribution in a marine snail, *Parcanassa* sp. The snails are carnivorous; they bury themselves in the mud as the tide goes out and emerge to crawl around on the surface of the mud as the tide comes in. They were counted in quadrats one foot square (sec. 10.1). The Poisson series having the same mean as the observations is shown in the third column. The sum of the fourth column gives χ^2_1. Columns 2 and 3 seem to agree quite closely; this is confirmed by the small value for χ^2 ($P = 0{\cdot}2$) and the value of 1·13 for the ratio of variance to mean which is close to unity. Apparently the snails were distributed over this area in a way that did not differ significantly from what might have been expected if they had been scattered over it at random.

Table 3.04 shows a similar set of measurements for the casts of a marine worm of the family Eunicidae. The worms live in burrows in the mud just above the level of low tide. They were counted in quadrats one foot square. The large value for χ^2 ($P < 0{\cdot}001$) indicates a significant

TABLE 3.03

The distribution generated by counting the number of *Parcanassa* in quadrats one foot square compared with the Poisson series

Snails per quadrat (x)	*No. of quadrats* (f)	*Poisson series*	$\frac{(O-E)^2}{E}$
0	228	223·32	0·127
1	64	64·45	0·003
2	6	9·25	
3	0		
4	0	1·98	1·591
5	1		
6+	0		
Total	299	299	1·721

$\frac{s^2}{\bar{x}}$ $\frac{0·3057}{0·2709} = 1·13$

$P(\chi^2)$ 0·2

TABLE 3.04

The distribution generated by counting the numbers of worms (Eunicidae) in quadrats one foot square compared with the Poisson series

Worms per quadrat (x)	*Number of quadrats* (f)	*Poisson series*	$\frac{(O-E)^2}{E}$
0	309	253·96	11·93
1	241	291·42	8·72
2	138	167·21	5·10
3	68	63·95	0·26
4	30	18·34	7·41
5	10	4·21	
6	3	0·81	
7	0		16·10
8	0	0·16	
9	1		
Total	800	800	49·52

$\frac{s^2}{\bar{x}}$ $\frac{1·590}{1·148} = 1·39$

$P(\chi_4^2)$ <0·001

departure from the Poisson series; and the value of 1·39 for the ratio of variance to mean indicates that the observed figures were more variable than the Poisson series; there were too many quadrats with 0 and more than 3 worms and not enough quadrats with 1 or 2 worms in them. Such a pronounced 'clumping' is unlikely ($P < 0{\cdot}001$) to have occurred from a random scattering of the worms. Some other hypothesis is needed (see sec. 3.32).

3.32 *The negative binomial*

Results like those in Table 3.03 in which the variance is no greater than the mean are rather rare in nature even when one chooses what seems to be an exceedingly uniform area for making the measurements. It is much more usual to find that the variance is greater than the mean, as in Table 3.04. This is what might have been expected because no area is so uniform that all parts of it are likely to be equally attractive or productive. Even if the area is uniform the animals may be gregarious or possess some other pattern of behaviour that leads to aggregation. There are a number of mathematical expressions that may be fitted to distributions in which the variance is significantly larger than the mean. Bliss (1953) suggested that the negative binomial might be one of the most useful because it is flexible; it can be fitted to a wide range of distributions ranging from those with variance equal to the mean to those in which the variance exceeds the mean by a large factor. The frequency distribution that is called the negative binomial is got by expanding the expression:

$$(q - p)^{-k}$$

where $q + p = 1$; $p = \frac{m}{k}$; m is the mean number of animals per quadrat and k is a positive exponent. When k is small – i.e. as it approaches zero – the negative binomial approaches a Poisson series. With moderate and large values for k the negative binomial becomes a series with variance larger than its mean.

In nature a large variance is the reflection of 'clumping': too many quadrats contain large numbers of animals with a consequent increase in the number containing none or very few and a decrease in the number of quadrats containing a moderate number of animals. Excessive 'clumping' might arise in a number of ways and it is possible to imagine a number of models to explain it. Bliss (1953) described two models that seem likely to occur rather frequently in nature: both fit the theory of the negative binomial.

According to one model, the negative binomial may be regarded as being compounded from a number of Poisson series in which the means vary in such a way that they are distributed like χ^2. For example, consider Cellier's (1958) data for the numbers of codlin moth larvae in apples that had been sprayed with different insecticides. There were 7 treatments including the control which was not sprayed at all. There were 12 trees sprayed the same way in each treatment. The fruit was harvested and stored in boxes for counting. The apples from each tree were not recorded separately. The counting was done by taking the apples at haphazard from the boxes in lots of 100 and recording the number that contained a codlin larva. If the trees had been scored separately this method of sampling would probably have generated a distribution for each tree that did not differ significantly from a Poisson series. But the trees were not scored separately, so the total sample was likely to represent the sum of a series of Poissons. The means of the different series were likely to be different because the trees had been sprayed differently and were therefore likely to have different numbers of codlin in the apples harvested from them. Consequently the results from this experiment make a likely model for the theoretical negative binomial that may be generated by compounding a number of Poisson series with different means. The results are set out in Table 3.05. The

TABLE 3.05

The distribution generated by counting codlin larvae in samples of 100 apples compared with the Poisson series and the negative binomial

No. of larvae per 100 apples	*No. of samples*	*Poisson series*	*Negative binomial*
0	300	214	301
1	159	231	159
2	83	125	83
3	48	45	43
4	16	12	22
5	15	3	11
6+	10	1	12
Total	631	631	631
$P(\chi^2)$		<0.000	0.4

second column gives the observed frequencies. The third and fourth columns give the frequencies for the Poisson and the negative binomial respectively. At the bottom of the table the values for P, taken from the table of χ^2, indicate the goodness of fit. The difference between the observed figures and the Poisson series were highly significant, but the negative binomial fitted the observed figures closely.

According to the second model a negative binomial may be generated by a distribution that is 'contagious' as Student (1919) pointed out. 'If the presence of one individual in a division increases the chance of other individuals falling into that division, a negative binomial will fit best, but if it decreases the chance, a positive binomial.' Gregarious animals

TABLE 3.06

The numbers of the gregarious isopod, *Trachelipus rathkei,* under boards; the observed distribution compared with the Poisson series and the negative binomial*

No. per board	*Observed*	*Poisson*	*Negative binomial*
0	28	4·5	30·2
1	28	14·9	21·6
2	14	24·5	16·2
3	11	26·9	12·3
4	8	22·3	9·4
5	11	14·7	7·3
6	2	8·0	5·6
7	3	3·8	4·3
8	3	1·6	3·4
9	3	0·6	2·6
10	3		2·0
11	2		1·6
12	0		1·2
13	1	0·2	0·9
14	2		0·7
15	1		0·6
16	0		0·4
17+	2		1·7
Total	122	122	122
$\frac{s^2}{\bar{x}}$	$\frac{15{\cdot}058}{3{\cdot}295} = 4{\cdot}6$		
$P(\chi^2)$		$<0{\cdot}000$	0·34

* After Cole (1946).

might be expected to be distributed 'contagiously' in the sense required for Student's model. Cole (1946) studied the pattern of distribution of the gregarious isopod, *Trachelipus rathkei,* by counting the numbers under 122 boards that he had placed on the ground. The results are set out in Table 3.06. The variance exceeds the mean in the ratio of 4 : 6, and the discrepancies between the observations and the Poisson series are large and significant. But the negative binomial fits the observations closely.

With the codlin larvae (Table 3.05) we had independent evidence from prior knowledge of the nature of the experiment and the methods of sampling, that the distribution was likely to be compounded of a number of distributions with variable means each resembling a Poisson series. With the isopods (Table 3.06) we had independent evidence from prior knowledge of the gregariousness of the isopods that the distribution was likely to be 'contagious' in the sense defined by Student, and therefore likely to resemble a negative binomial. But it is not permissible to argue in the opposite direction. A significant discrepancy from the Poisson series disproves the hypothesis of random scatter but agreement with a theoretical series does not in itself permit any inference about the biology of the particular animals that were counted.

Consider, in this connexion, Garman's data for European red mite on apple leaves, quoted by Bliss (1953). The sample of 150 leaves were picked from 7 trees. The figures are set out in Table 3.07. The variance is nearly twice as large as the mean and a large χ^2 indicates a highly significant ($P < 0{\cdot}001$) discrepancy between the observations and the Poisson series. But the negative binomial fits the observations closely ($P = 0{\cdot}3$).

TABLE 3.07

Garman's data for the relative abundance of red mites on apple leaves. The observed distribution is compared with the Poisson series and the negative binomial*

No. of mites per leaf	*No. of leaves observed*	*Negative binomial*
0	70	69·49
1	38	37·60
2	17	20·10
3	10	10·70
4	9	5·69
5	3	3·02
6	2	1·60
7	1	0·85
8+	0	0·95
Total	150	150
$\frac{s^2}{\bar{x}}$	1·98	
$P(\chi^2_2)$		0·3

* After Bliss (1953).

There are several ways in which this particular distribution may have been generated. (*a*) The density of the population of mites probably varied from tree to tree. But the mites on each tree may have been distributed at random. In these circumstances the distribution got by counting the mites on 150 leaves taken at random from 7 trees would represent the sum of a series of Poissons with different means. (*b*) If the trees had been sprayed with mitecide it is likely that some leaves would carry a heavier dose of spray than others. The lightly sprayed would be more likely to harbour mites than the heavily sprayed ones, thus fulfilling Student's condition for contagion. (*c*) Another sort of contagion might develop if the mites had multiplied locally after originally colonizing the trees, perhaps at random. Any of these models might generate a negative binomial. But the mere fact that the negative binomial fits the observations does not help us to choose between the several explanations. If we wish to explain the good fit we must seek independent biological evidence.

3.33 *The influence of 'aggregation' on the accuracy of sampling for density*

When quadrats (sec. 3.22) are used to estimate the density of a population the size of the sample (the number of animals counted) that is required for a particular level of accuracy depends on the variance. Larger samples will be needed for the same degree of accuracy when the variance is large than when it is small. For certain sorts of 'contagion' the size of the sample required for a particular degree of accuracy increases roughly in proportion to the average size of a 'cluster'. A knowledge of the theoretical expression that provides a fit for the observations may be helpful not only in defining the accuracy of sampling, but also in indicating a suitable transformation for further statistical analysis. But this problem is beyond the scope of this book.

When the method of capture, marking, release and recapture is used the accuracy of the results depends less on the way that the animals are distributed than on the thoroughness with which the marked ones mix with the unmarked after the marked ones have been released. Consequently the method of marking, release and recapture may give more accurate results than the method of quadrats when the aggregation is pronounced. (See sec. 10.2.)

3.4 FURTHER READING

The student who wishes to follow up the matter of 'density' might consult the original papers that are mentioned by Andrewartha and

Birch (1954), Chapter 13 at the beginning of each sub-section dealing with the ecology of the 12 species that are discussed in that chapter. In most instances the authors have described their original methods for estimating the densities of the populations that they studied.

More advanced methods than the rather elementary ones that I have described in section 3.23 have been developed for analysing the results of capture, marking, release and recapture. They give better estimates of birth-rates, death-rates and densities; also expressions for the variance of r_{+}, r_{-} and a have been worked out which allow the precision of estimates to be known. These methods may be found in papers by Jackson (1940, 1944, 1948), Bailey (1951, 1952), Leslie (1952) and Leslie and Chitty (1951).

For the student who wishes to follow up the discussion in sections 3.31 and 3.32 Bliss (1953) contains a useful bibliography.

CHAPTER IV
Dispersal

4.0 INTRODUCTION

Broadly speaking, animals have three ways of dispersing – by drifting with currents of air or water, by swimming, walking or flying, and by clinging to some moving object; the moving object may be some sort of animal, as a tick may cling to a sheep (sec. 6.11), or it may be some article of commerce, as a grain weevil is dispersed in a sack of grain.

4.1 DISPERSAL BY DRIFTING

Drifting cannot always be clearly distinguished from swimming or flying because many species may get a start this way or even maintain themselves afloat or aloft by their own efforts while being carried by the current. But it is worth making the distinction because this is the chief method of dispersal for so many small arthropods that are serious pests, and for most of the small animals that live in the sea; also many of the larger animals in the sea have planktonic larvae that disperse in this way.

4.11 *Dispersal of the bean aphis* Aphis fabae

The bean aphis lives by sucking sap from the bean, *Vicia faba*, especially from young plants. It is a serious pest because it sometimes becomes very abundant and retards the growth of the crop. Aphids as a group do a lot of damage to horticultural crops not only by sucking sap but also by spreading virus diseases. Therefore a knowledge of the dispersal of aphids is relevant to many important practical problems in horticulture. The principles that apply to the dispersal of *A. fabae* are likely to apply to the dispersal not only of other sorts of aphids but also to many other small animals that drift in the air. Johnson and his colleagues at Rothamsted have made a thorough study of the dispersal of *A. fabae* (C. G. Johnson, 1950b; 1952a, b; 1954; 1955; 1956; 1957; Johnson and Taylor, 1957; Taylor, 1957).

There may be a number of generations of *A. fabae* during the year.

At different times of the year the population may be composed predominantly of wingless parthenogenetic females, winged parthenogenetic females or winged sexual forms. It is the winged parthenogenetic females that are most important in dispersal. There may be three main 'waves' of dispersal, in the spring, summer and autumn.

Normally the winged parthenogenetic female flies away from the plant on which it has been reared before it begins to reproduce; if it subsequently alights on a suitable plant it will feed; in these circumstances it begins very shortly to produce young. At the same time the muscles of the wings are autolysed and their substance contributes to the rapid production of more progeny. The first young may appear within half a day after the female has alighted on a suitable plant and started to feed. When the aphids were kept in the dark they did not fly. Embryos began to develop in such aphids just as in those that had flown. But the muscles did not autolyse and the development of the embryos was delayed; several days elapsed before the first young were produced. The autolysis of the wing muscles is controlled by a hormone. The secretion of the hormone may be delayed in the aphids that are not allowed to fly (B. Johnson, 1953). These observations show how the innate tendency to disperse is nicely geared to the physiology of reproduction through an endocrine mechanism; there is nothing casual or superficial about this behaviour; it is an essential part of the animal's life.

Most of the aphids fly upwards as they leave the plant. Taylor (1957) measured, at half-hourly intervals, the number of *A. fabae* flying immediately above a crop of beans on which there was a dense population with many of them moulting to the winged stage. He found the greatest numbers when the light intensity exceeded 1,000 lumens. They became rapidly fewer as the intensity of the light became less than this. By an hour before sunset, when the intensity of the light was about 10 lumens, there were scarcely any aphids leaving the beans. Johnson and Taylor (1957) made similar observations with respect to temperature. They concluded that most individuals required a temperature of 17·5°C or higher before they would fly; but a few would fly when the temperature was as low as 16·5°C and there were a few that did not fly until the temperature reached 19·5°C.

Taylor (1957) used the term 'teneral' to define the development that takes place between the final moult, when the aphid appears with fully developed wings, and the stage when it is ready to fly away. The duration of the teneral period is determined by temperature. At 16°C the teneral period lasted about 48 hours; at 25°C it lasted for about 8 hours; the shortest period observed in nature was about 6 hours. Even in warm weather when the temperature in the crop may exceed 25°C for several hours during the middle of the day, aphids that moult during the latter part of the day are unlikely to be ready to fly before nightfall; but many of them may have completed teneral development by morning. Thus a relatively large proportion of the population is likely to become ready for flight during the period when flight is inhibited by darkness or cold. Most of them take off during the middle of the morning as the temperature reaches the threshold for flight (16·5°–19·5°C). This gives rise to the characteristic peak of activity in the mid-morning. There is also a less well-defined peak in the afternoon which has not been explained.

Apart from the influence of temperature on the mid-morning peak the number of aphids in the air immediately above the beans seemed not to be consistently related to temperature, humidity or wind. About all that could be said about the relationship of weather to 'flight-activity' was that most of the aphids began their flight in windy weather because most of the weather was windy.

The number of aphids in the air immediately above the crop depends on the density of the population on the beans, the stage of development of the individuals that constitute the population and their activity in taking flight. In order to find out what happened to them after they had left the plant, Johnson measured the number in the air at different levels from 4 feet up to 2,000 feet. He used the suction traps that I mentioned in section 3.1. He also measured the gradient in atmospheric temperature because it seemed likely that the movement of aphids to and from the upper levels might depend on the turbulence of the air.

During a typical summer's day the following regime seemed to hold. Most of the aphids were more than 100 feet from the ground. Usually very few were found in the air at night; the few that were present were usually at great heights. During the morning the numbers at all levels increased. About an hour after the mid-morning 'peak' of activity at the level of the crop aphids would be present in substantial numbers up to 2,000 feet or more. During the day high numbers were maintained at all levels; but towards evening they decreased rapidly leaving only a few at high altitudes. During spring and autumn much the same sequence of events was observed except that there were usually fewer aphids at great heights and relatively more at the lower levels. There were some exceptions. Sometimes during summer the upper levels would contain relatively few aphids; sometimes the aphids disappeared quite abruptly from the upper levels quite early in the day.

All these observations could be related to the turbulence of the air. The turbulence is not easily measured directly but it may be inferred from measurements of the gradient in the temperature. If the air at the higher altitudes becomes colder than that near the ground it becomes unstable and turbulent. Conversely, a gradient in the opposite direction leads to stability and scarcely any turbulence. When the air was turbulent the aphids were carried up to great heights. The greater the turbulence the greater the proportion of aphids at the higher levels. Some were also carried down but on balance those going up outnumbered those going down, and high numbers were maintained at all levels up to several thousand feet. When stability was restored to the air, usually in the evening the upward movement ceased and the aphids settled out of the air rapidly. In general the air is unstable during the day and becomes stable at night; it is relatively more stable during autumn and spring than during summer, which is the period of greatest instability. These changes are related to diurnal and seasonal changes in insolation and radiation of heat from the ground. The stability of the air, as indicated by the gradient in temperature, was closely related to Johnson's measurements of the numbers of aphids in the air.

It is not yet known how long an aphid is likely to stay in the air or how far it is likely to travel. But now that the principles have been established these details will doubtless be filled in. Perhaps it is best to conclude this summary with a statement of the present theory as Johnson (1955) described it.

> Aphids do not die merely by being carried into the upper air: these aerial populations are alive. And though many of the individuals may never find host plants and survive to continue the race, nevertheless this kind of dispersal, and the behaviour which accompanies it (upward flight at take-off and subsequent flight in relation to host acceptability) is of a kind to which the whole life and evolution of the species is attuned. There is no evidence that justifies the belief that it is a few which behave in some other way which are the only ones which need to be considered.
>
> We see the whole migrating population being physically mixed by turbulent diffusion high into the air and descending with variable frequency: and those finding the right place as belonging to this multitude. . . .
>
> The double peak of density at crop level is often reflected in a double peak of total aerial numbers. This suggests that flight is then

rather limited in duration to something less than the duration of a density peak – i.e. about 4 hours – and that on those occasions the first flight makes a major contribution to the daily numbers flying.

However, though the aerial population in general descends by nightfall, it is common to observe extremely low densities high in the sky, and these may persist through the night and add up to a vast number of insects; and though the majority of the day fliers may sometimes travel no farther than a wind of 2 or 3 hours' duration will carry them, the low residual densities appear to indicate that a few persistent fliers may travel much farther.

Johnson's conclusions were especially interesting because they contradicted the 'theory' that had been widely accepted. This 'theory' may be summarized briefly. Aphids cannot fly against a wind of more than 2 or 3 feet per second and do not take flight readily except in calm air or when there is a very light breeze. They fly more readily when the humidity is high than when it is low. Therefore dispersal is largely restricted to calm moist weather.

Johnson (1954) examined this 'theory' and found that it was largely based on experiments that had been done by Davies (1935, 1936, 1939) who found that certain aphids behaved like this when tested in the laboratory. The results of Davies' experiments have not been doubted, but the logical superstructure that was built upon them was faulty. In the light of present knowledge it is easy to see what the faults were. It is true that in nature, as in the laboratory, aphids take off more readily in a gentle breeze than in a stiff one but a moderate breeze does not inhibit them altogether. In a crop of beans there are sheltered places where the air is relatively calm even during a strong wind. Strong winds are usually turbulent so that even on a windy day there are temporary lulls which provide opportunities for the aphids to take flight. There is much more windy weather than calm weather so that in the aggregate most aphids begin their flights in windy weather. The errors in the logical deductions that were made from the results of Davies' experiments would not have done much harm if the conclusions had been recognized as a model and used to stimulate further experiments with the aphids in the field.

But this did not happen, for nearly 20 years progress was delayed while the 'facts' about the natural dispersal of aphids continued to be deduced from the results of Davies' laboratory experiments. Certain observations were made and certain experiments were done during this period but their results made little impact on the ideas that were so firmly entrenched. The 'theory' was indeed no theory but merely the conclusion of a conceptual model masquerading as theory (sec. 9.3 *et seq.*, especially sec. 9.33).

4.12 *Dispersal of wingless insects, mites and spiders*

The principles that have been established for the dispersal of *A. fabae* are likely to be helpful with many other species that use this method of dispersal. Coad (1931) dragged a net behind an aeroplane and estimated that the air between 50 feet and 14,000 feet above one square mile near Tallulah, Louisiana, contained 25 million insects and other small animals. Most of them were in the first 1,000 feet but some were found at all levels up to 14,000. The larger, heavier ones and the stronger fliers were closer to the ground; the smaller ones, the weaker fliers and the wingless ones constituted most of the 'catch' at the higher altitudes.

There is convincing evidence that the scale insect *Eriococcus orariensis* which recently spread dramatically through the indigenous stands of *Leptospermum scoparium* in New Zealand was originally blown from Australia, a distance of more than 1,000 miles across ocean. This species occurs on *Melaleuca* in Australia; it is most unlikely to have been carried accidentally to New Zealand either by tourists or by traders. Moreover the original outbreak was traced back to a lonely and inaccessible mountain valley where it was most unlikely to have been carried by any means other than wind. The speed with which it subsequently spread in New Zealand is also good evidence of the effectiveness of its dispersal by wind (Hoy, 1954).

Duffey (1956) studied the dispersal of 9 species of linophiid spiders that are common in the Wytham estate near Oxford. They all have one generation a year. Some species disperse in the spring, others in the autumn. But with all of them, dispersal by drifting with the wind is a normal part of their behaviour which recurs at the same stage of the life-cycle in every generation.

I have been impressed by the ubiquity of the small sap-sucking mites *Tetranychus* on the plants in my garden. The mites breed on a variety of herbaceous garden plants and weeds. It is most striking that they usually manage to colonize a seedling within a few days of its appearance above ground. It seems that they are, through most of the summer, floating in the air in such large numbers as to make the early discovery of any new plant a near certainty.

4.13 *The dispersal of locusts*

Locusts can fly powerfully as any one who has tried to catch them with a net may testify. Yet, strangely enough, most of the large-scale dispersal of locusts is done by drifting with winds at high altitudes. When a swarm of locusts has been resting for the night they remain on the ground next morning, more or less quiescent, until the temperature rises above a certain threshold which may vary with the turbulence of the air. Usually they fly more readily when the air is more turbulent. So they are often carried upward by ascending currents of air until they are several thousands of feet above the ground (Rainey and Waloff, 1951). A locust that is flying close to the ground usually seeks shelter if the wind becomes strong. But a locust that is high enough to be out of sight of the ground no longer perceives changes in the direction or the velocity of the wind and it may remain aloft for a long time and be blown hundreds of miles.

Waloff (1946a) traced the path of one swarm of *Schistocerca gregaria* from Morocco to Portugal. Waloff (1946b) analysed many records of the movements of swarms of *S. gregaria* in eastern Africa; and Rainey and Waloff (1948) did the same for the area around the Gulf of Aden. Most of the long migrations were the outcome of the locusts having been blown with the prevailing winds, especially those at altitudes of 2,000 to 3,000 feet. There was no evidence that the migrations led, except by chance, to favourable places for breeding; but when, in the course of their wanderings with the wind, the locusts arrived at a place that was moist enough, they would develop to sexual maturity and lay eggs.

4.14 *The dispersal of strong-flying insects by winds near the ground*

Coad (1931), see above, found that most of the large strong-flying insects in his 'catches' were caught at lower levels and it seems that the dispersal of many insects of this sort is regularly helped by surface winds. Barber (1939) found the fly *Euxesta stigmatias* breeding in an isolated field of sweet corn in Florida in circumstances that strongly indicated that they had been blown in from the West Indies. Dorst and Davis (1937) traced the dispersal of the beet leafhopper *Eutettix tenellus* from winter breeding grounds across 200 miles of desert (where there was no food for the leafhoppers except on the experimental plots used for trapping them) to the beet-growing areas of Utah. Felt (1925) pointed out that the moth *Heliothis armigera* rarely, if ever, overwinters north of southern Pennsylvania and the Ohio river, but it sometimes becomes abundant on sweet corn in southern Canada. Felt also listed a number of other species from the tropics or the sub-tropics which were unable to survive a northern winter, yet from time to time were found breeding in Canada during summer. It is likely, in all these instances, that the animals had been blown north by winds but we do not know whether they travelled in the upper air; it seems more likely that they may have been blown by winds near the ground.

4.15 *The dispersal of marine species that drift with currents*

Most of the smaller animals in the sea disperse by drifting. Many species have planktonic larvae that are especially adopted for drifting with the current. Some members of the permanent plankton are so characteristic of particular currents that they have been called 'current indicators' (Russell, 1939).

Certain barnacles and oysters that live in tidal rivers and estuaries have planktonic larvae whose behaviour shows special adaptations which allow the population to persist in these places despite the fact that the net flow of water is outwards towards the ocean. Like the aphids, they make use of turbulence. The denser water near the bottom tends to have a net flow inland. Barnacles, as they approach the settling stage, tend to swim down and congregate near the bottom (Bousfield in Clarke, 1954). The larvae of oysters, as they get a little older, tend to cling to the bottom during the ebb tide and rise into the water during the flood tide (Carricker, 1951).

The dispersal of the European eel involves both drifting and swimming at different stages in the life-cycle (Schmidt, 1925). The spawning of the eel has not been described. The small flat planktonic larvae can first be found floating in the water in an area south-east of Bermuda. They drift with the Gulf Stream. After about 3 years they reach the coast of Europe or North Africa. They metamorphose to elvers which swim up the rivers. After several years' development in fresh water the mature eels leave the rivers and return to the same area that they came from as larvae. It is not known how they find their way back to the same place. It seems likely that they do it by swimming rather than drifting.

4.2 DISPERSAL BY SWIMMING, WALKING OR FLYING

Certain butterflies, fish, birds and whales are known to 'migrate' great distances – hundreds or even thousands of miles. The animals may undertake these long journeys quite regularly as a characteristic feature of their lives. They may resemble the eel (sec. 4.15) in that the whole population, or the whole breeding population, moves from one extreme of the distribution to the other. Many birds, like the eels, breed only at one end of the distribution but, unlike the eels, the same individuals may make the journey repeatedly twice every year (Mathews, 1955; Hochbaum, 1955). The butterfly *Danaus plexippus* breeds in Ontario during the summer; each autumn they migrate southwards. They overwinter in southern North America. Beall (1946) shows that the immigrants that returned to Ontario the next summer were not the same individuals that had left in the autumn but were almost certainly their progeny. Several other species of butterflies in Europe behave similarly. Migration is a large and specialized subject so I shall not discuss it in this book.

4.21 *The measurement of dispersal*

Dobzhansky and Wright (1943, 1947) described a method for measuring the rate of dispersal of animals that disperse chiefly under their own power. They used *Drosophila pseudoobscura*. They arranged a series of traps in two straight lines that bisected each other at right angles (Fig. 4.01). They then set free about 4,000 *Drosophila* at the place where the

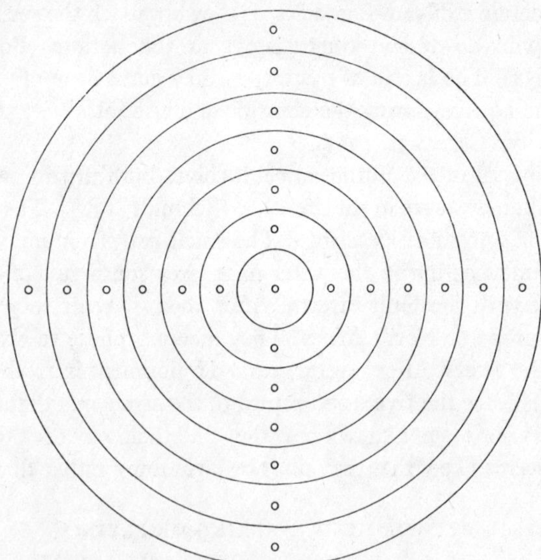

Fig. 4.01. The distribution of traps in a cross as used by Dobzhansky and Wright to measure the dispersal of *Drosophila*. Theoretically each set of 4 traps measures the density of the population in an annulus.

two lines of traps intersected. Subsequently they counted (and released) the *Drosophila* at each trap on each of 9 consecutive days. The number counted at the central trap was taken as an estimate of the population in the central circle and the numbers counted at the four traps in each annulus were taken as an estimate of the population in the annulus. The mean density in each annulus may be regarded as the frequency with which individuals occur at that distance from the centre (point of origin).

If we imagine that these frequencies have been arbitrarily divided between two directions from the centre (instead of four), the results may be set out in the conventional form of a frequency distribution whose mean has been arbitrarily set at zero (the position of the central trap, Fig. 4.02). Irrespective of whether 4 or some other number of traps was used to measure the density of the population in an annulus, the variance of r (the distance travelled by an animal) may be estimated as follows.

Let $p =$ the distance of any trap from the centre.

Let $d =$ the diameter of the central circle and the width of each annulus. For convenience let us work in units of d, i.e. let us call $d = 1$.

Let $\lambda = \dfrac{\text{number of animals in unit area}}{\text{number of animals in a trap}}$.

Let $f =$ number of animals in a trap in an annulus.

Let $\bar{f} =$ mean of f.

Let $c =$ number of animals in a trap in the central circle.

Let $\bar{c} =$ mean of c.

The area of central circle $= \pi\left(\dfrac{d}{2}\right)^2$.

The number of animals in this area $= \dfrac{\pi d^2 \lambda \bar{c}}{4}$.

The area of an annulus of radius $r = 2\pi rd$.

The number of animals in this area $= 2\pi rd\lambda \bar{f}$.

To obtain the variance we proceed as usual to find the sum of squared deviations from the origin and divide this by the total number of observations. In this instance:

$$\text{The variance of } r \text{ (i.e. } s^2) = \frac{S2\pi rd\lambda \bar{f}r^2}{S2\pi rd\lambda \bar{f} + \pi\dfrac{d^2}{4}\lambda\bar{c}}$$

$$\text{(where } d = 1) \quad = \frac{S\bar{f}r^3}{S\bar{f}r + \dfrac{\bar{c}}{8}}$$

The results of Dobzhansky and Wright's experiment are shown in Table 4.01. The variance for each of the 9 days is shown in the third column; the variance was first calculated in trap-units and then converted to square metres.

TABLE 4.01

The dispersal of *Drosophila pseudoobscura*: variance in square metres, dispersal in metres*

Day	*Temp.* °C (T)	*Variance* s^2 *in* m^2	*Daily increment in* s^2 (i)	*Standard deviation* × 0·67 *in m.* (s × 0·67)
1	13	3,200		373
2	19	5,000	1,800	474
3	19	7,400	2,400	561
4	15	7,200	−200	569
5	13	6,100	−1,100	523
6	18	8,000	1,900	599
7	17	8,700	700	625
8	17	11,900	3,200	731
9	15	10,500	−1,400	687

* After Dobzhansky and Wright (1947).

An increase from 3,200 to 10,500 in 9 days gives an average of 811 per day. But the daily increments in variance that were actually observed varied from −1,400 (an unrealistic value almost certainly due to sampling errors) to 3,200. The variability in the daily increments in variance was related to temperature. Dobzhansky and Wright found a significant correlation between T and i (sec. 5.15). There is no evidence that i was greater at the beginning of the experiment than at the end. Apparently the rate of dispersal was independent of the degree of crowding. On the other hand, the presence of a few high frequencies near the tail of the distribution in Figure 4.02 shows that the flies were heterogeneous with respect to 'dispersiveness'; i.e. the sample included a few individuals that dispersed more actively than the majority. This feature may be characteristic of many species because it has been observed in most of those that have been studied carefully.

In any normal curve half of the population is separated from the origin by a distance equal to or greater than 0·67 × the standard deviation. Thus the fifth column in Table 4.01 shows for each day the distance away from the centre that half the flies would have reached or passed. This estimate depends on the theory of the normal curve. In the present instance the observed frequencies differ rather strongly from a normal curve because of the heterogeneity that I mentioned above. So the figures in column 5 are likely to contain substantial errors. So long as this limitation is remembered no harm is done in presenting them as a rough measure of the dispersal of *Drosophila pseudoobscura*.

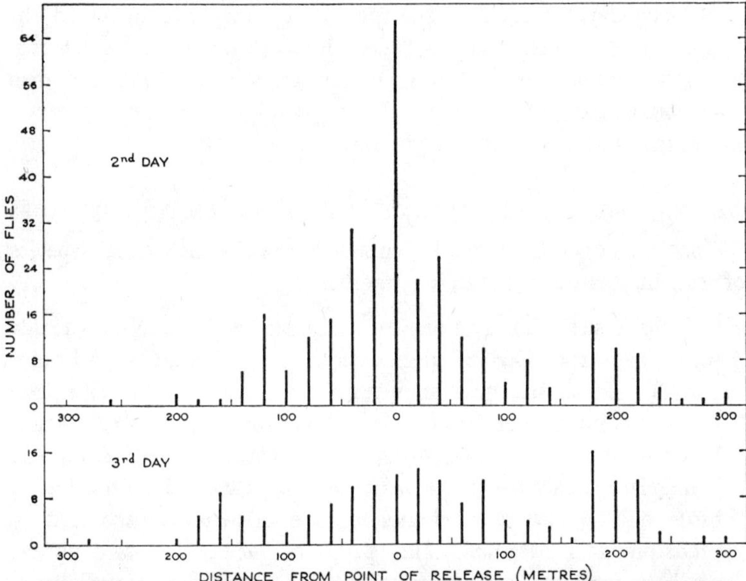

Fig. 4.02. The distribution of marked flies (*Drosophila*) on two successive days in one of Dobzhansky's experiments. See text for explanation. The unit for the abscissae is 20 metres, which was the distance between traps (Fig. 4.01). The ordinates show the mean number of flies in a trap. (After Andrewartha and Birch, 1954)

Dobzhansky and Wright used one trap in the central circle and four in each of the annuli. This design can be improved by using several traps in the central circle and varying the number of traps in the annuli in proportion to their area. The accuracy of the estimate of density depends on the number of animals caught not on the number of traps used. Therefore it is better to use more traps where the density is low (sec. 3.33). Moreover the frequency in each annulus is multiplied by the square of the distance from the centre. So errors in the estimates for outer annuli are magnified more than those for inner annuli. Experiments of this design that can be done in class are described in section

4.3 DISPERSAL BY CLINGING

This is the usual way of dispersal for parasites; for some, like the sheep tick *Ixodes ricinus*, it is the only way; for others like some of the parasitic worms, it may be supplemented by a limited amount of movement, or by other means, during the free-living stages.

Modern developments in transport and trade have increased the opportunities for dispersal afforded to the wide variety of animals that may attach themselves to or conceal themselves in the things that men transport in the course of trade. Elton (1958) described a number of 'invasions' that have occurred this way.

4.4 THE SPECIAL IMPORTANCE OF DISPERSAL IN INSECT PREDATORS

Cockerell (1934) described the circumstances in which certain species of scale insects were kept in low numbers by predators.

> I recall some observations on Coccidae made in New Mexico many years ago. Certain species occur on the mesquite and other shrubs which exist in great abundance over many thousands of square miles of country. Yet the coccids are only found in isolated patches here and there. They are destroyed by their natural enemies, but the young larvae can be blown by the winds, or carried on the feet of birds, and so start new colonies which flourish until discovered by predators and parasites. This game of hide-and-seek doubtless results in frequent local extermination, but the species are sufficiently widespread to survive in parts of their range, and so continue indefinitely.

Flanders (1947) made the same point more generally. He analysed the effectiveness of hymenopterous 'parasites' as agencies of 'biological control'. These terms are used as the economic entomologist uses them: 'effectiveness' relates to ability to reduce the abundance of the prey to the level where it is no longer a pest; 'parasite' is the special sort of predator whose larva lives inside the prey and destroys it. Flanders first pointed out:

> In many cases it is known that the population level of a host species is determined by the host searching capacity of the free-living gravid females of the parasitic insect. The actual level attained is determined by those attributes of the female parasite that contribute to its host-searching capacity.

He then described the circumstances in which the searching goes on:

> When a host is under biological control, it has a characteristic 'colonial' or 'spotted' distribution (Smith, 1939). The host population consists of groups of individuals, which fluctuate in density independently of one another. One group may be exterminated by a parasite, but in the meantime some host individuals will have migrated and established new groups. The parasite must therefore follow its host, and so must equal it in powers of locomotion. . . . The female of *Metaphycus helvolus* is very active and apparently travels considerable distances between ovipositions. Observations indicate that populations of *M. helvolus* may disperse a distance of 25 miles in one year (Secs. 9.12, 9.2; Figs. 9.03, 9.08).

There is fairly general agreement among students of biological control that a predator should be able to 'out-breed' and 'out-disperse' its prey if it is to have a reasonable chance of reducing its prey to low numbers (sec. 6.211).

4.5 THE GENERAL IMPORTANCE OF DISPERSAL

Elton (1927) wrote: 'The idea with which we have to start is that animal dispersal is on the whole a rather quiet, humdrum process, and that it is taking place all the time as a result of the normal life of the animals.' If you look again at the examples that I have discussed in this chapter you will see that these animals have, as a normal part of their lives, patterns of behaviour that tend to speed the processes of dispersal and make them more effective. One can often observe in nature circumstances that might be expected to favour the evolution of such patterns of behaviour.

Food and suitable places to live are often scattered 'patchily' and perhaps sparsely through a terrain that is otherwise inhospitable. Moreover the pattern changes from time to time perhaps as a result of ecological succession, perhaps from catastrophe such as flood or drought or fire, or perhaps from the industry of men; or the activities of the animals themselves may speed the process of succession – whatever the cause, places that were suitable for occupation today may have gone by tomorrow, and new ones may have arisen where there was none before. In these circumstances it may be an advantage to a species to breed individuals that readily seek a new place to live, even while the old one is still quite favourable. The passage from Cockerell that I quoted in section 4.4 shows how high powers of dispersal may be an advantage to a species that has to live with a predator that presses heavily upon it, tending to annihilate the local populations that it finds. The following passage which I quote from Nicholson (1947) shows how high 'dispersiveness' is also an advantage to a herbivorous insect that presses heavily on its food-plant, tending to annihilate local populations when it finds them. This passage summarizes the sequence of events that happened after the moth *Cactoblastis cactorum* had been introduced into Queensland with the hope that it would reduce the abundance of prickly pear, *Opuntia*, which was at that time a widespread and serious weed. The venture was a striking success.

> After the introduction of this moth into Australia it increased in abundance at a terrific rate, quickly destroying the prickly pear over large areas around the points of liberation. After some time the point was reached at which very little prickly pear remained, whereas there were in the country countless millions of moths that had bred on the prickly pear destroyed during the previous generation. Most of the caterpillars coming from the eggs of these moths died for simple lack of food, and most of the remaining prickly pear was destroyed at the same time. But this did not produce the complete eradication of the prickly pear. By chance, or because of some favourable circumstance, here and there prickly pear plants survived. Similarly, even at this time some few of the moths succeeded in finding odd plants on which to lay their eggs, although because of the great reduction in numbers of *Cactoblastis* in the previous generation, many plants were not found. The end result, which still persists, is that prickly pear is scattered in small isolated groups with wide intervals between them. In a few of these *Cactoblastis* is still to be found, and these are generally doomed to complete destruction because *Cactoblastis* is able to increase rapidly in numbers upon them. In other groups of prickly pear, which have not so far been found by *Cactoblastis*, the pear tends to spread; but sooner or later it is found by the moth, and the destruction of these groups is achieved shortly afterwards. In the meantime seed is scattered in new places, so maintaining the existence of prickly pear. Consequently, within Queensland as a whole, there is a low density of prickly pear, and a low density of *Cactoblastis*, which vary little from year to year; but at the same time there is continual fluctuation in space.

Relatively few animals press so heavily on their food or are pressed on so heavily by predators as those mentioned in the passages quoted from Cockerell and Nicholson. But there are many other causes for the temporariness and the 'fluctuation in space' of local populations. Animals need to be constantly dispersing as a normal part of their lives – and those that happen to evolve special patterns of behaviour which speed the processes of dispersal are likely to be favoured by natural selection.

4.6 FURTHER READING

Andrewartha and Birch (1954, Chap. 5). I have scarcely mentioned in this chapter the long-range dispersal across large ecological barriers. Dispersal on this scale is especially interesting in relation to the study of Zoogeography. It is discussed by Darlington (1938, 1956) and Elton (1958).

CHAPTER V

Components of Environment: Weather

5.0 INTRODUCTION

The component of environment that I have called 'weather' relates to the particular degree of temperature, moisture, light, wind, etc. that may be measured in the precise place where the animal is living. This is not the same as the conventional weather that is recorded by the meteorologist because meteorological measurements are standardized: in Australia, for example, temperature and humidity are measured in a 'Stevenson screen' which is a louvred white wooden box placed 4 feet from the ground. 'Weather', in the ecological context, may be modified in a number of ways that are purposely excluded from the records that are taken by meteorologists. In particular, animals that crowd together may influence the weather experienced by one another, especially if there are many of them. When there is a cold wind blowing, a sheep in a large closely huddled mob may be in a warmer place than a sheep that is by itself or one of a small mob (sec. 6.11).

The weather experienced by an animal may also be modified by the place where it is living (sec. 8.2). And weather may modify the influence that food, predators, pathogens and other components of environment may have on an animal's chance to survive and multiply. It will be best to leave the discussion of these interactions for later chapters. In this chapter I shall discuss only the more direct influence of weather as a component of environment.

5.1 TEMPERATURE

5.11 *The influence of temperature on speed of development*

At 25°C *Thrips imaginis* completes its development from egg to adult in about 9 days and the weevil *Calandra oryzae* requires about 35 days at the same temperature. A convention has grown up among zoologists to use the reciprocals of such measurements as a measure of the speed of development of the animal. Thus it is said that the speed of development of *Calandra* at 25°C is 2·9% per day; but the speed of development of *Thrips* at 25°C is 11·1% per day. It is immediately clear that these two measures of 'speed' are not to be compared with each other because they refer to proportions of quite different quantities. The usefulness of this method is limited to comparing the speed of the same developmental process (i.e. the same stage in the life-cycle of members of the same species) at different temperatures. It has been widely used for this purpose.

In nature, animals usually live in fluctuating temperatures. But it is difficult to do experiments with fluctuating temperatures; also the results of such experiments are difficult to interpret. So I shall first discuss the results of experiments in which temperature was kept constant and then (in sec. 5.112) discuss the consequences of fluctuating temperatures.

5.111 The speed of development at constant temperatures

Figure 5.01 gives the results of Powsner's (1935) experiments with the eggs of *Drosophila melanogaster*. The points on the descending curve represent the raw data as it came from the experiment; they show the number of hours that the eggs took to complete their development at each temperature. The points on the ascending curve are the reciprocals of the original data and represent the conventional measure of the speed of development.

The relationship between temperature and speed of development has been described by the particular sigmoid curve that is called the logistic curve. The general formula for the logistic curve is:

$$\frac{1}{Y} = \frac{K}{1 + e^{a+bt}}$$

where Y is the duration of development at temperature t, and K, a and b are empirical constants. For an ascending curve the constant b is negative.

In Figure 5.01 the empirical points lie closely along the calculated curve up to 30°C. Above 30°C the harmful influence of high temperature causes the observed points to fall below the theoretical curve. Between the limits 17° to 25°C the curve is rather flat and a straight line would fit the points in this range fairly well.

Until Davidson (1944) pointed out the superior merits of the logistic curve the straight line had been widely used. The straight line is the

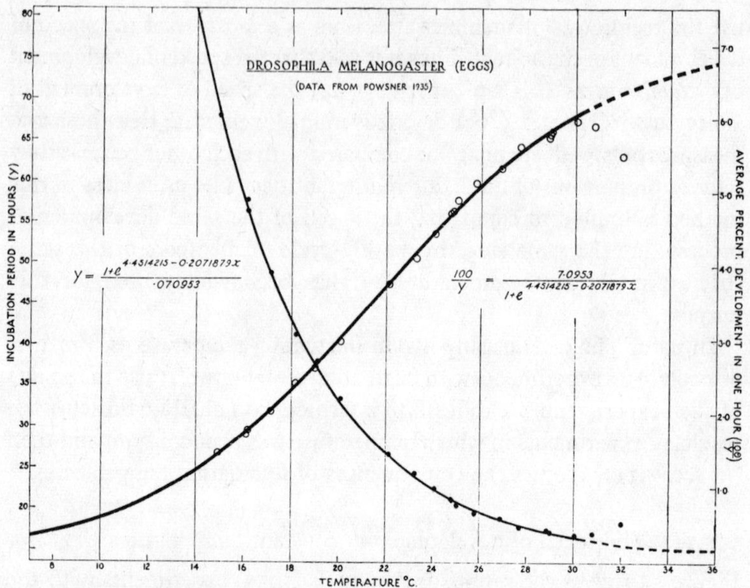

Fig. 5.01. The relationship between 'speed of development' and temperature of the eggs of *Drosophila melanogaster*. The hours required to complete the egg-stage are plotted against temperature (*descending curve*) and the reciprocal of this time is plotted against temperature (*ascending curve*). The points plotted on both curves are observed points. The ascending curve is a logistic curve and the descending one is its reciprocal, as indicated by the formulae shown on the diagram. (After Davidson, 1944)

reciprocal of the hyperbola. The general formula for the hyperbola is:

$$Y = \frac{K}{t - d}$$

where Y is the duration of development at temperature t, and K and d are empirical constants.

For the reciprocal of Y, which is the conventional measure for speed of development, this equation can be written:

$$\frac{1}{Y} = k + bt$$

where Y and t have the same meanings and k is a new constant equal to $-bd$. This is the equation of a straight line and d can easily be shown to be the value of t when $\frac{1}{Y}$ becomes zero. It has been customary, therefore, to call d the 'threshold of development' and (this leading to the error that Weiss complained about – sec. 1.3) to imagine that all development ceases at the so-called 'threshold of development'. But it is now well known, because any well-conducted experiment will demonstrate it, that the temperature–time curve approximates to the hyperbola for only a very short range of temperature; and the concept of *d* as the 'threshold of development' is unrealistic. The only merit in the straight line is that it is easily calculated; for data that have not been gathered precisely it may be good enough for a first approximation (sec. 11.1, exp. 11.12).

5.112 The speed of development at fluctuating temperatures

Ludwig and Cable (1933) exposed the pupae of *Drosophila* on alternate days to the following pairs of temperatures 21° and 29°C, 23° and 28°C and 15° and 25°C. In each instance the duration of development was the same as what would have been expected from experiments done at constant temperatures. Birch reached the same conclusion from experiments with eggs of the grasshopper *Austroicetes cruciata*. He used the following pairs of temperatures: 19° and 31°C and 23° and 27°C. These results are typical of others that might be quoted. It seems that short-term fluctuations in temperature, provided that they do not include extremes, may safely be considered as equivalent to constant temperatures. This is not to say that the average speed of development is the same as it would be at that constant temperature which is equal to the mean of the fluctuating temperatures. This would imply that the relationship between temperature and speed of development is linear, which it is not.

When the fluctuations include extremes outside the favourable range, it is necessary to consider the following possibilities: (*a*) that short exposures to an extreme temperature may impair the animal's competence to develop at a favourable temperature and (*b*) that healthy development may be possible during short exposures to extreme temperatures that would be harmful if they were prolonged. These points were discussed by Andrewartha and Birch (1954, sec. 6.232). We also discussed the anomalies that may occur when diapause is present, which I do not have space to mention in this book.

5.12 *The lethal influence of temperature*

The lethal influence of low temperature has been the subject of much study, especially by entomologists in Europe and North America because they have been interested in the survival-rates during winter of various insect pests. The lethal influence of high temperature has received much less attention. With terrestrial animals the difficulty may be to measure the influence of heat independently of dryness. In nature harmful high temperatures are more likely to occur in deserts or in warm temperature zones where the summer is arid. In these circumstances dryness may be more important than heat. High temperature may be more important for certain species of fish that live in shallow lakes or rivers in temperate climates. There have been a number of experiments done on fish to determine the range of temperature within which they can live and the influence of acclimatization on this range.

The influence of lethal temperature cannot be measured by the same direct methods that are appropriate for measuring the influence of temperature on the speed of development. For example one of Johnson's experiments with the eggs of *Cimex* which I describe below was designed to find out how long the eggs would live at 4·2°C. While the eggs remained at 4·2°C there was no way of telling by inspection which were dead and which were still alive. In order to discover this they had to be removed from the cold and incubated in the warmth. Then it was possible to tell what proportion of the sample had died during exposure to cold but it was not possible to tell for any one egg what 'dose' of cold had been required to kill it. In other words, there was no way of measuring directly the susceptibility of an egg to cold, and consequently no direct method of estimating the mean susceptibility of the eggs to cold as we could, for example, estimate their mean weight or the mean duration of their life-cycle. The way around this difficulty is to design an experiment that allows us to estimate the dose that is required to kill just half of the population. This provides an estimate of the median which, for a normal curve, is equivalent to the mean (sec. 11.2).

If the problem is to measure the duration of exposure to a particular temperature that is required to kill half the population, we begin by exposing a number, say six samples of 50 or more animals each, to this temperature for varying periods. If the problem is to measure the temperature that is required to kill half the population when they are exposed to it for a particular period, then all the samples are exposed for the same period but to different temperatures. In either case the treatments are chosen with the expectation that the least severe will kill a small proportion of the sample, the most severe will kill about 80% and the remainder will straddle the 50% mark as nicely as may be. It may be necessary to keep one of the samples as a control, i.e. it is not exposed to the harmful temperature at all. As each sample completes its treatment it is removed to a favourable temperature, kept for a suitable period, several days or more, and the number of deaths recorded.

If these numbers, expressed as proportions, are plotted against the treatment (or dose) they will fall along a sigmoid curve, which is the familiar 'dose-mortality' curve of the toxicologist. This curve is also the one that can be got by progressively summing the ordinates of the ordinary frequency-polygon. Provided that the frequency-distribution of deaths against 'dose' is normal or nearly so, the sigmoid 'dose-mortality' curve becomes a straight line when the proportions are transformed to probits (Finney, 1947). If the distribution is not normal it may be made so by transforming the dose to logarithms or some other appropriate scale.

Once the straight line has been calculated, it is easy to read off, by linear interpolation, the dose (temperature or duration of exposure) that would be expected to kill any proportion of the population. The dose required to kill 50% corresponds to the mean dose that we could have calculated directly if we were able to write down by direct inspection of each animal the dose just required to kill it. We may sometimes require to know the dose required to kill some other proportion of the population. There is no difficulty in this, but it must be remembered that, unless the size of the sample is increased, the estimates will be less precise. To estimate the dose required to kill 95% the sample requires to be 3 times, and for 99% 10 times, as large as that required to estimate, with the same degree of accuracy, the dose required to kill 50%. It is usual, in writing about probit analysis to abbreviate 'the lethal dose required to kill just 50 per cent' into 'L.D. 50'.

Johnson (1940) used the probit method to measure the influence of low temperature on the survival-rate in eggs of the bed bug *Cimex*. This paper should be studied in the original by the student who is interested in technique.

In one experiment, eggs of a uniform age (0–24 hours at 23°C) were exposed to three temperatures, 4·2°, 9·8° and 11·7°C for periods ranging from 9 to 50 days. At the conclusion of each exposure the eggs were incubated at 23°C. Those that did not hatch were considered to be dead. The results are set out in Table 5.01. The death-rate is expressed both as a per cent and as a probit. The former is transformed to the latter simply by reference to the appropriate table in Finney (1947) or Fisher and Yates (1948). Table 5.01 also gives the values for *b*, L.D. 50 and L.D. 99·99. The first is the coefficient for linear regression of probit on dose; it measures the slope of the probit line and inversely the

TABLE 5.01

Death-rate of eggs of *Cimex* exposed for various periods to 4·2°, 9·8° and 11·7°C*

4·2			*9·8*			*11·7*		
Exposure (days)	*Death-rate*		*Ex-posure (days)*	*Death-rate*		*Ex-posure (days)*	*Death-rate*	
	%	*Probit*		%	*Probit*		%	*Probit*
9	4·2	3·27	7	10·1	3·72	7	9·4	3·7
16	16·9	4·04	14	7·5	3·56	14	9·7	3·7
23	52·5	5·06	21	19·9	4·15	21	36·7	4·7
30	86·3	6·09	28	65·8	5·41	29	59·2	5·2
35	94·5	6·60	35	86·3	6·09	35	69·4	5·5
40	91·8	6·39	42	100·0	8·72	42	90·7	6·3
44	94·5	6·60				46	94·0	6·6
50	100·0	8·72				49	96·2	6·8

	4·2	9·8	11·7
b	0·1139 (0·000151†)	0·1273 (0·000241)	0·0800 (0·000058)
L.D. 50	22·9 days (1·120)	25·8 days (0·774)	26·7 days (1·409)
L.D. 99·99	55·6 days	55·0 days	73·3 days

* After Johnson (1940).
† Figures in parenthesis are variances.

variance. The L.D. 50 and L.D. 99·99 are the estimated durations of the exposures required to kill 50% and 99·99% respectively. The variances for *b* and L.D. 50 are also included because these quantities are necessary for the calculation of the statistical significance of the results.

Figure 5.02A shows the death-rate, in per cent, plotted against the duration of exposure, and Figure 5.02B shows the same data plotted as probits (see also Figs. 11.21*a*, 11.22*a*). The advantages of transforming the data to a scale in which the regression becomes linear are that it enables us to make a precise estimate of the exposure required to kill any proportion of the population; and an exact significance can be attributed to the observed differences. Figure 5.02B indicates that the probit lines for 4·2° and 9·8°C are nearly parallel and that both slope more steeply than the probit line for 11·7°C. On the other hand, if we compare the co-ordinates of the abscissae for probit 5 (i.e. the value for L.D. 50) we find that 9·8° and 11·7°C come close together whereas 4·2°C is well away to the left. These two statistics, *b* and L.D. 50, are the two most instructive ones to extract from the probit line. The L.D. 50 is the most reliable measure of the animal's capacity to withstand exposure to a given temperature, and *b* is the best comprehensive measure of the relative toxicity of short and long exposures. A large value for *b*, i.e. a steep probit line, indicates that there is little 'scatter' between the exposures that kill few and many; conversely a small value for *b*, i.e. a gently sloping probit line, indicates a wide margin between the exposures required to produce these extremes in the death-rate (sec. 11.2).

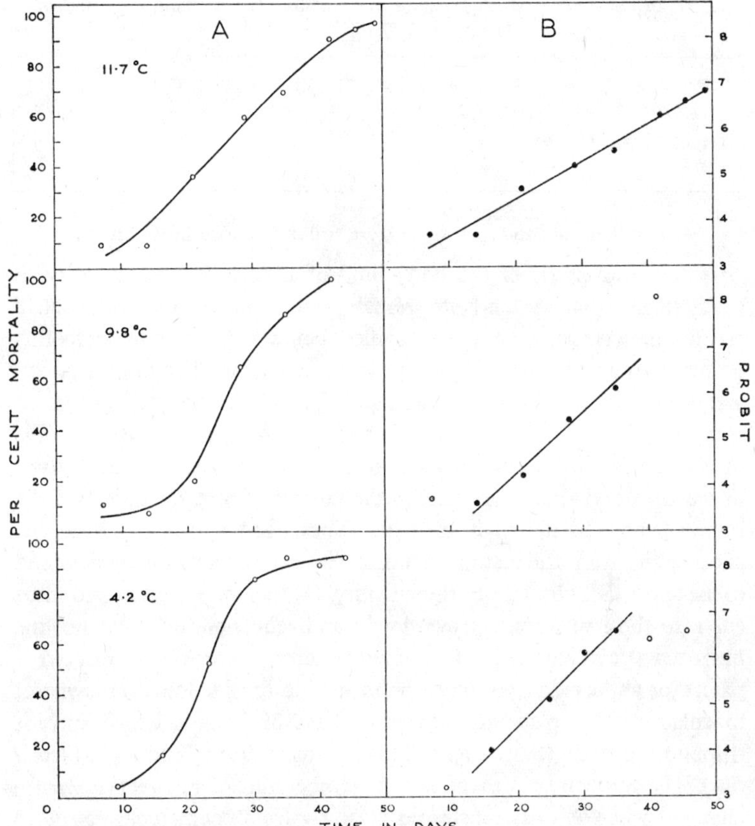

Fig. 5.02. The death-rate for eggs of *Cimex* exposed to 4·2°, 9·8° or 11·7°C for a varying number of days. A, death-rate expressed as a per cent is plotted against time; B, by transforming from per cent to probit the sigmoid curve becomes a straight line. (Data from Johnson, 1940)

The sawfly *Cephus cinctus* is indigenous to the prairies of western Canada where it has become a pest of wheat. It spends the winter as a prepupa inside the stem of the wheat where it may be exposed to temperatures below 0°C. Salt (1950) kept a sample of 18 prepupae at

−20°C for 120 days; he recorded the day on which each one became frozen. The results are given in Table 5.02. It seems that cold-hardy insects like *Cephus* are not likely to be killed by prolonged exposure to cold unless they freeze, and not always then.

TABLE 5.02

Time in days required for freezing to occur in overwintering prepupae of *Cephus**

Day	1	2	3	4	5	8	13	15	16	28	33	42	66	71†
No. of prepupae freezing	0	0	2	1	1	1	1	1	1	1	1	2	1	1

* Data from Salt (1950).
† On the 120th day there were still 4 prepupae unfrozen.

It is characteristic of the body fluid of animals (especially of cold-hardy ones) that it tends to remain unfrozen, in a 'super-cooled' condition, when the animal is chilled below 0°C. The supercooling occurs because the water is present in a finely divided condition associated with colloids. For this reason, and also because one organ or tissue may be quite different from another in this respect, the first outcome of chilling is almost certainly the formation of small isolated crystals of ice here and there throughout the animal; doubtless there will be more in some tissues than in others. These, being insulated from one another and from the rest of the unfrozen water by colloids (which tend to become dehydrated in the vicinity of the ice crystals and thus enhance the insulation), grow slowly and tend not to 'inoculate' the unfrozen supercooled water. Sooner or later, however, a threshold is passed or an 'accident' occurs; the insulation breaks down somewhere; inoculation takes place and sets off a wave of freezing which spreads through the body (Salt, 1950). This is illustrated in Figure 5.03 (*solid line*). The curve illustrates the way the temperature of an insect (*Ephestia*) changed when it was transferred abruptly from room temperature to −30°C. The temperature fell steeply and smoothly to −24·3°C, when it abruptly increased to −18·6°C and then proceeded to fall again. Supercooling ceased at −24·3°; the wave of crystallization that followed liberated enough heat to raise the temperature by 5·7°C. Then, as this heat was absorbed the temperature began to fall again. The temperature at which supercooling ceased is called the 'undercooling point'; the rise in temperature is the 'rebound'; and the temperature to which the rebound rises is called the 'freezing point'.

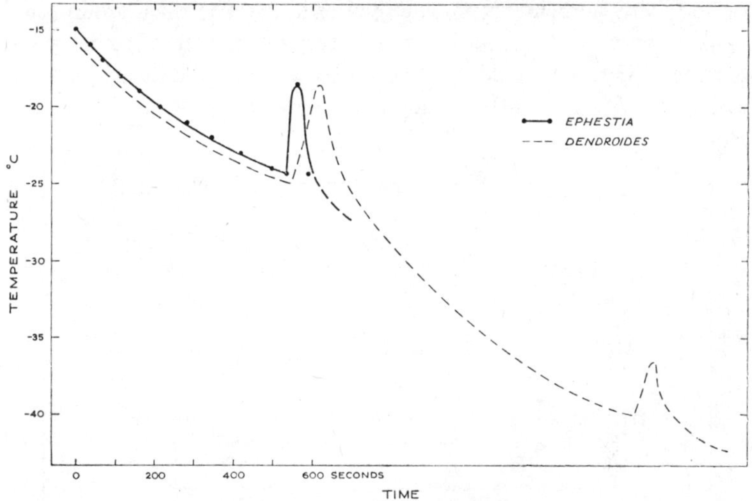

Fig. 5.03. Characteristic curves for the temperature of an insect that has been abruptly transferred from room temperature to some low temperature, say −40°C. The complete line is characteristic of most species which die after the first rebound. The broken line is characteristic of the few very hardy species that can survive the first rebound. (After Andrewartha and Birch, 1954)

Kozhantchikov recognized that insects could be divided into three groups according to their response to low temperature. (1) The ones in the first group cannot survive for any considerable period if the temperature falls below the lower limit of the range favourable for normal development. They cannot become dormant at low temperatures; they must either develop or die. This group is made up of (*a*) species that live in or originated from tropical or sub-tropical climates, e.g. *Cimex*, *Calandra* and *Locusta*; there are many of these species in which no stage in the life-cycle is capable of becoming dormant if the temperature falls too low for development; (*b*) species from temperate climates in which there is a certain stage in the life-cycle that is adapted to survive the winter but all other stages resemble the tropical forms in lacking the capacity to become dormant at moderate or low temperatures. (2) The second group in Kozhantchikov's classification comprises the forms that can become 'quiescent' as Shelford (1927) defined this term. That is, while remaining competent, at any time, to develop if they are placed in the warmth, they may nevertheless survive, healthy but inactive, at temperatures too low for development. Contrary to popular belief this group is very small; in searching the literature it is difficult to find many authentic representatives of it. (3) The third group contains the special over-wintering stage of the species adapted to the cold-temperate climates. Cold hardiness and dormancy is nearly always associated with a condition which is known as diapause (Andrewartha, 1952).

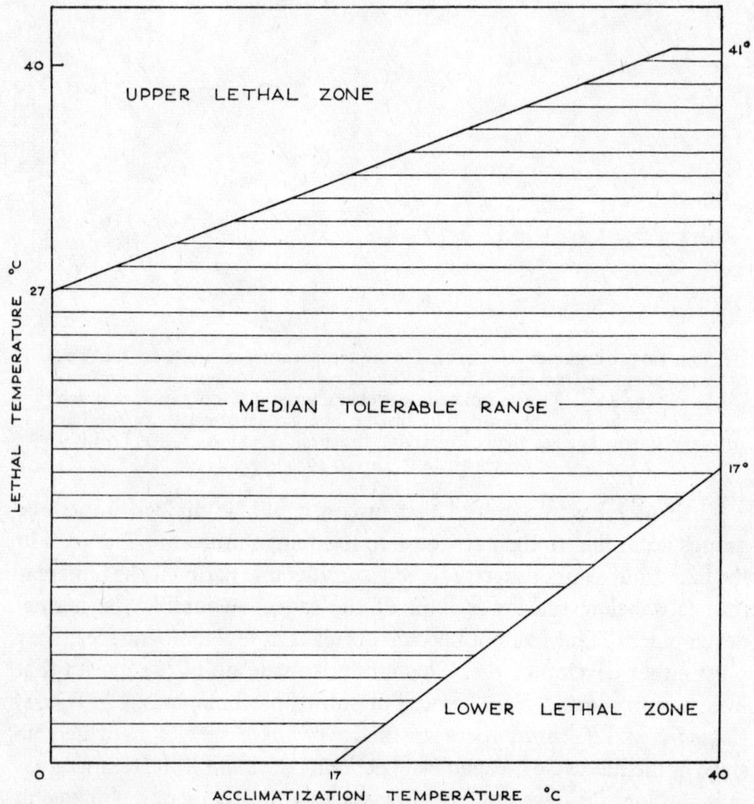

Fig. 5.04. The median tolerable range of temperature for goldfish. Notice how the limits of the tolerable range may be influenced by the temperature at which the fish had been living. (After Andrewartha and Birch, 1954)

5.13 *The limits of the tolerable zone*

If an animal dies promptly when it is exposed to an extremely high or low temperature there may be no difficulty in attributing its death to the exposure to extremes of temperature. At less extreme temperature death may take longer but it may still be recognized as being largely caused by exposure to unfavourable temperatures. But as we move farther and farther away from the extremes we come to a zone of moderate temperature where it is not reasonable to attribute any lethal influence to temperature. Fry (1947) called the two limits to this median tolerable zone the 'incipient lethal low temperature' and the 'incipient lethal high temperature' and he defined the extremes outside these limits as the 'upper lethal zone' and the 'lower lethal zone' respectively. Can we discover the limits of the median tolerable zone by experiment?

When we come to plan the experiment, we find that it cannot be done precisely because, as the temperature becomes less extreme, the exposure becomes so long that we may not be able to decide whether the animal died from old age or from some harmful influence of temperature. In their experiments with young trout, Fry *et al.* (1946) used periods that varied from 20 to 80 hours depending on the circumstances: the periods were long enough to justify the assumption that any fish that was still alive at the end of the experiment would have been able to survive an indefinitely long exposure to that particular temperature.

5.131 The influence of acclimatization on the limits of the tolerable zone

Fry *et al.* (1942) measured the limits of the tolerable zone for goldfish that had previously been acclimatized to temperatures ranging from just above 0° up to 40°C. The results are shown in Figure 5.04. Both the lower and the upper incipient lethal temperature were higher for fish that had been living at higher temperature; and the distance between the two limits was narrower for fish that had been acclimatized to a high temperature. The incipient lethal high temperature could not be raised above 41°C by acclimatization. Taking into account the full influence of acclimatization the lowest incipient lethal low temperature for goldfish may be about 0°C and the highest incipient lethal high temperature about 40°C.

The ability to extend the limits of the tolerable zone by acclimatization may be valuable to fish that live in shallow water. Huntsman (1946) showed that extremes of temperature sometimes killed many fish in Ontario. Brett (1944) collected samples of bullhead (*Ameiurus*) from Lake Opeongo at intervals during the year and measured the high temperature required to kill half the sample during an exposure of 12 hours. Figure 5.05 shows that the seasonal trends in 'heat-hardiness' were large and striking.

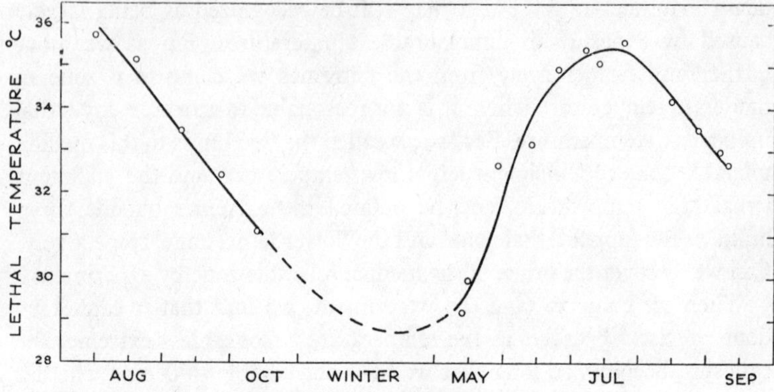

Fig. 5.05. The seasonal trend in the 'heat-resistance' of the bullhead *Ameiurus*. The ordinate gives the L.D. 50 for an exposure of 12 hours to high temperature. The fish clearly have become acclimatized to the prevailing temperature of each season. (After Brett, 1944)

5.14 *Behaviour in a gradient of temperature*

Doudoroff (1938) placed a number of fish *Gisella nigricans* in a tank of water in which the temperature varied from 20° at one end to 30°C at the other. They arranged themselves with the frequencies shown in Figure 5.06. Most of the fish congregated near 26°C; ordinates from the abscissa at 25° and 28° enclose about 75% of the area under the curve: more than 60% of the fish stayed between the limits of 25° and 27°. This band of preferred temperature has been called the 'preferendum'.

The locust *Chortoicetes terminifera* has a preferendum at about 42°C. Clark (1947, 1949) studied the behaviour of the young nymphs of this locust in a place where the initial population reached about 2,000 per square yard in the most densely populated parts. In the morning, as the sun warmed the ground, the young locusts followed the gradients in temperature until they came to congregate in the places where the temperature approached more closely to 42°C.

The results of Doudoroff's experiments with *Gisella* may be attributed without any reasonable doubt to temperature because temperature was the only variable in this experiment. And it seems likely that the behaviour of *Chortoicetes* may be properly attributed to temperature. But experiments of this sort are often difficult to interpret because of interactions between temperature and other components of weather. In water the concentration of dissolved oxygen is likely to vary with temperature. In air changes in absolute humidity are slow and small

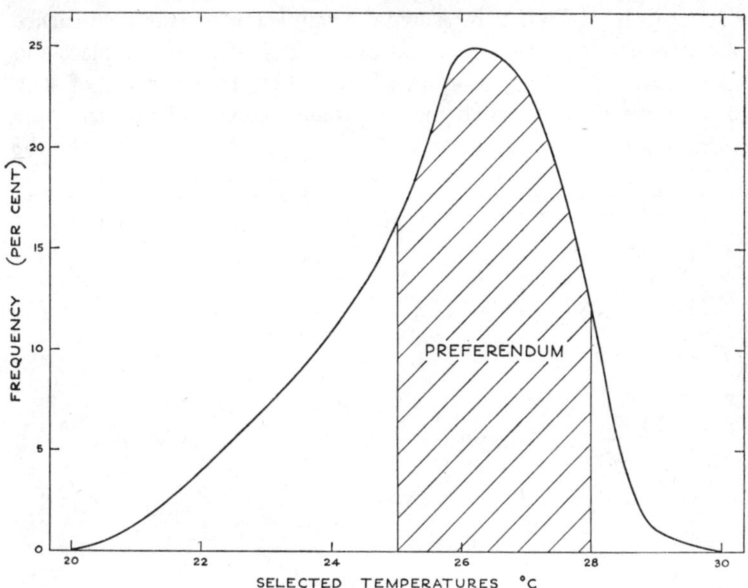

Fig. 5.06. The 'preferendum' for the fish *Gisella nigricans*. The ordinates indicate the per cent of the fish choosing each temperature. About 75% of the fish congregated in the range indicated by the shading. (After Andrewartha and Birch, 1954)

relative to temperature; consequently, fluctuations in the evaporative power of the atmosphere, whether measured directly as 'evaporation' or indirectly as 'saturation deficit', follow fluctuations in temperature (Fig. 5.07).

5.141 The influence of acclimatization on behaviour in a gradient

The fish that were used in Doudoroff's experiment (Fig. 5.06) had been living at about 15°C before the experiment began. In another experiment fish were 'acclimatized' at temperatures ranging from 10° to 35°C before being tested. The result was that the preferendum varied directly with the temperature at which the fish had been living (Fig. 5.08). But there was an upper limit beyond which acclimatization had no further influence. Fry (1947) called this upper limit to the preferred temperature the *final preferendum*. It is indicated in Figure 5.08 by an arrow.

5.15 *Influence of temperature on activity and dispersal*

In section 4.21 I mentioned Dobzhansky and Wright's experiment with

Drosophila in which they measured the daily increments in the variance of a population of *Drosophila* that had been set free in one place and allowed to move freely away in all directions. In one series of such experiments they measured the daily temperature and calculated the linear regression of 'increase in variance' on temperature. On the

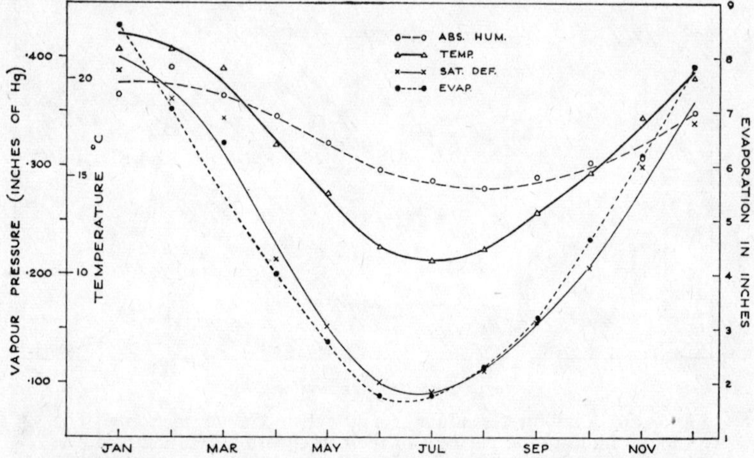

Fig. 5.07. The annual trends in atmospheric temperature and humidity at the Waite Institute, Adelaide. The curves are based on the mean daily records for 23 years. Note that changes in absolute humidity are slight and gradual compared to the more abrupt changes in saturation deficiency and evaporation, which are closely related to temperature and absolute humidity. (After Andrewartha and Birch, 1954)

average the 'increase in variance' was greater by 924 square metres for each 1°F increase in temperature. But the relationship was not truly linear. The flies moved very little when the temperature was below 15°C. When the temperature in the evening (which is the time when the flies are most active) was about 20°C the flies travelled, on the average, about 120 metres; with temperatures around 25°C they probably travelled about 200 metres in a day.

Davidson and Andrewartha (1948a, b) used the statistical method of partial regression to measure the influence of temperature on the dispersal of *Thrips imaginis*. This is a small winged insect about 2 mm. long. The thrips were strongly attracted to roses and these were used as 'traps' (sec. 3.1). The thrips were breeding in a variety of other flowers in and around the garden. But they are characteristically restless and rarely stay in one place for long. A proportion of those flying in the garden during the day settled in the roses and some of them were found there when we collected our daily sample in the morning. These records were kept for the months August–December (southern spring) for 14 years. A characteristic series of records is shown in Figure 7.01 which represents the daily numbers of thrips per rose during 1932 (Evans, 1932).

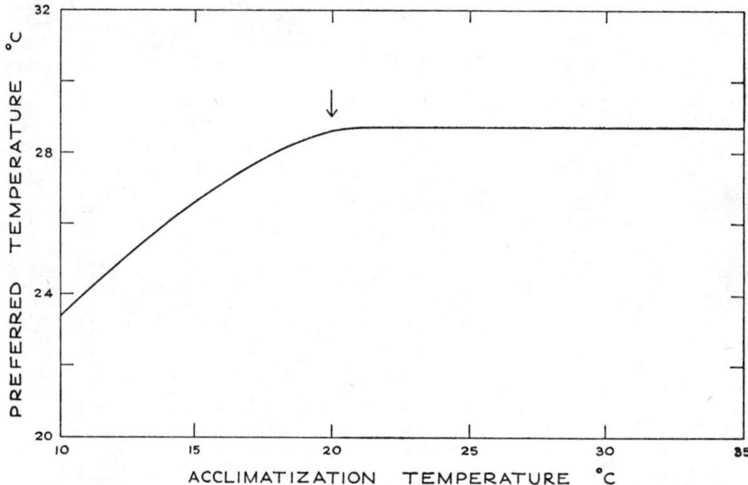

Fig. 5.08. The 'final preferendum' temperature for the fish *Gisella nigricans*. For explanation see text. (After Andrewartha and Birch, 1954)

The seasonal trend and the enormous daily fluctuations about the trend shown for this year are characteristic of all the years for which records were kept. In order to analyse the data we first transformed the numbers to logarithms; this was done chiefly because proportional changes in numbers were more instructive than additive changes. It was then possible to account for the systematic trend of numbers with time by fitting empirically a curvilinear regression of the form:

$$Y = a + bx + cx^2 + dx^3 \qquad (5.1)$$

where Y represents the logarithm of the number of thrips and x represents days. Figure 5.09 for 1935 is typical. The systematic trend that this curve describes is due chiefly to the natural increase of the population, but it may also be partly caused by trends in the weather. The important point is that the curve has accounted for the trend with time, whatever may be its cause, so that the residual variance (represented by the departures of the observed points from the trend-line in Fig. 5.09) is independent of time. These departures are largely due to daily variations in the activity of the thrips moving into and out of the 'traps'. On a warm day more thrips moved out of the flowers in

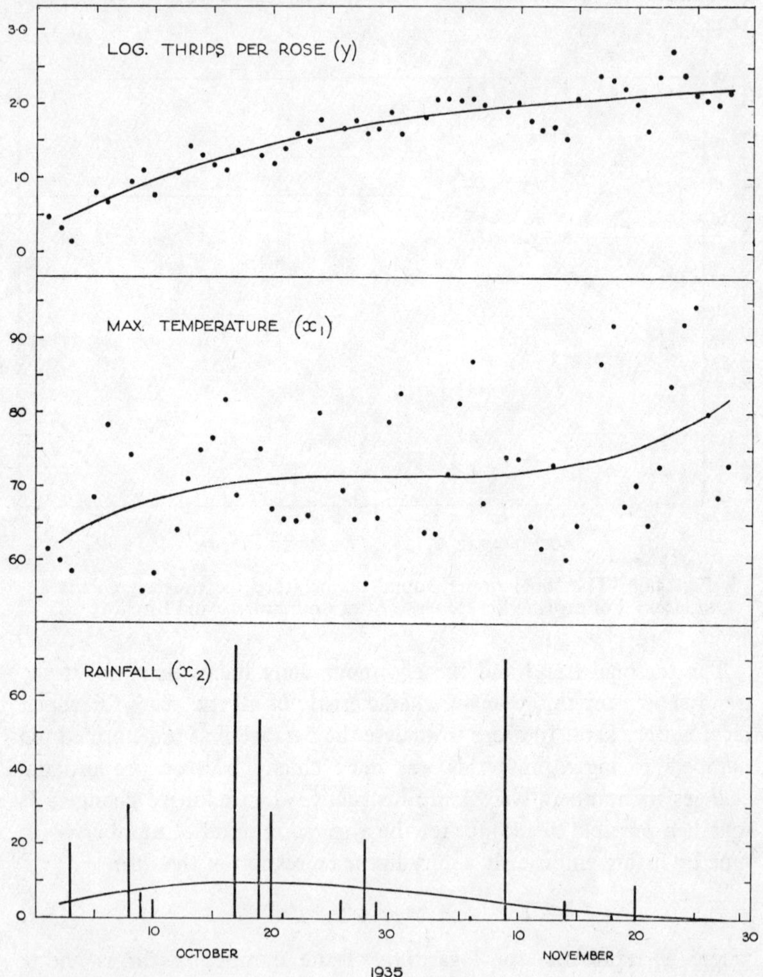

Fig. 5.09. Some of the data (for the year 1935) for equation 5.2. The smooth curves have in each instance been obtained by calculating the appropriate polynomial of the form indicated in equation 5.1. Temperature is measured in degrees Fahrenheit and rainfall in units of 0·01 inch. For further explanation see text. (After Davidson and Andrewartha, 1948b)

the surrounding fields, where they were breeding, and found their way into the newly opened roses, which were sampled daily.

The method of partial regression provides a way to measure the degree of association between the 'activity' of the thrips (as represented by deviations from the smooth curve) and whatever components of environment that may seem important. The final step in the analysis is to solve the equation for partial regression which may be written:

$$Y = b_1 x_1 + b_2 x_2 + b_3 x_3 \qquad (5.2)$$

where Y represents the departure of the observed data from the smooth curve in Fig. 5.09; $x_1, x_2, x_3, \ldots$ represent temperature, humidity, and other components of environment also expressed as departures from the smooth curves (Fig. 5.09); and $b_1, b_2, b_3, \ldots$ are the coefficients for partial regression such that b_1 measures the association of Y with x_1, independent of any relations that might exist between these two and any other variables in the equation.

In the present example it was found that the independent influence of temperature was highly significant and that, on the average, the numbers of thrips in the flowers on any one day increased by 5% for every increase of 1°C in the daily maximal atmospheric temperature.

The analysis also showed that the influence of temperature was the same whether there were few or many thrips in the population. Apparently the movement of the thrips to and from the breeding places to the roses was independent of the density of the population. It was not the outcome of jostling or any other manifestation of numbers; it was rather likely to have been the outcome of an innate tendency to disperse possessed equally by thrips living in solitude and those living in a crowd. This tendency may be influenced by several components of weather of which temperature is perhaps the most important (sec. 4.5).

5.16 *Adaptations to temperature*

'Acclimatization' refers to physiological changes that occur in an individual during its lifetime; they are usually reversible. 'Adaptation' refers to genetical differences that have arisen by mutation and selection. They are reversible only by the same evolutionary processes.

Moore (1939, 1942) compared the speed of development of the eggs of northern species of frogs (*Rana*) with those of southern species (Table 5.03). He used the time taken to develop from stage 3 to stage 20 as a criterion. These are arbitrary stages in the morphogenesis of amphibian embryos that were described by Pollister and Moore (1937). There is a

TABLE 5.03*

The duration of development of the eggs of four species of *Rana* at three different temperatures in relation to the temperature of the water where they usually breed

Species of Rana	*Northern limit of distribution (° latitude N)*	*Usual temp. of water at breeding time (°C)*	*Hours required to develop from stage 3 to stage 20*			*Ratio column (b) to (c)*
			12°C (a)	*16°C (b)*	*22°C (c)*	
R. sylvatica	67°	10°	205	115	59	1·95
R. pipiens	60°	12°	320	155	72	2·15
R. palustris	51°–55°	15°		170	80	2·12
R. clamitans	50°	25°		200	87	2·29

* After Moore (1942).

close relationship between the temperature of the water where the species usually breeds and the speed of development of the eggs at constant temperatures. Those from cold places developed more rapidly at low temperatures than those from warmer places. But the acceleration with rising temperature was greater for those from warm places; the most southern species *R. clamitans* developed from stage 3 to stage 20 in 45 hours at 33°C – a temperature that killed the other three species.

Similar adaptations to temperature may be observed with respect to the limits of the tolerable range. Brues (1939) found the larvae of certain Diptera living in thermal springs in Indonesia at temperatures up to 52°C. At the other extreme the beetle *Astagobius angustatus* lives in ice-grottos where the temperature ranges between −1·7° and 1°C (Allee *et al.*, 1949). But no individual species is known that can thrive over such a wide range as from 0° to 50°C. The range of temperature favourable to a particular species is related to the prevailing temperatures in the places where the animals usually live. Not only the average or usual temperature but also the range of temperature are reflected in the physiology of the animals. Figure 5.10 shows that for species from cold places, like *Dendronotus*, the favourable range is low and for those from hot places, like *Thermobia*, it is high. Also the limits of the favourable range are wider for terrestrial than for marine species, which reflects the greater variability of temperature on land than in the sea.

5.2 MOISTURE

About the 'wettest' place that an animal can live in is fresh water. About the 'driest' place for an aquatic animal is a rock-pool of sea-water that is drying up. The brine shrimp *Artemia* can live in such places and so too can the larvae of certain species of mosquito. Certain 'terrestrial' animals such as rotifers, tardigrades and nematodes live in the film of water that surrounds, for example, rotting vegetation; their places of

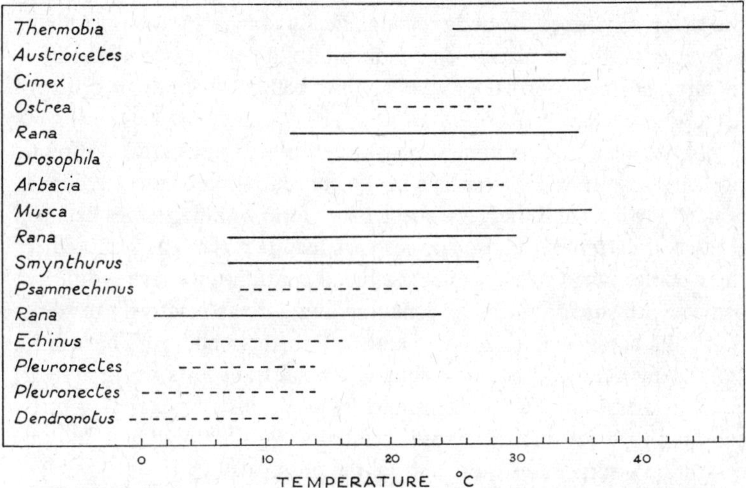

Fig. 5.10. The range of temperature favourable for the development of a number of aquatic (broken lines) and terrestrial or amphibious animals (solid lines). Note (*a*) that the ranges for the aquatic species are usually shorter than for the terrestrial ones and (*b*) the wide range between species in the limits of the favourable range. This is related to the temperatures that prevail in the places where the animals live. The three lines for *Rana* refer to three different species. (After Andrewartha and Birch, 1954)

living can scarcely be discriminated from those of aquatic animals living in fresh water. Animals that are truly terrestrial live in drier places than the 'driest' of aquatic places; some live in just about the driest places that one can imagine. The larva of the moth *Ephestia* can live in flour that has been dried in an oven at 105°C and stored in a desiccator over concentrated sulphuric acid. The kangaroo rat *Dipodomys* can live in the desert in Arizona with no water except what it can get by eating grain or seeds that are in equilibrium with the dry air of the desert.

The turbellarian *Gunda ulvae* can live either in sea-water or fresh stream-water provided there is enough calcium in it (Krogh, 1939). After *Gunda* had been in stream-water for an hour it had increased 1½ times in volume compared with its volume in sea-water; in stream-water that lacked calcium it swelled until it burst. Certain other aquatic animals are like *Gunda*, but most species respond to changes in the 'wetness' of the place where they are living (i.e. to changes in the osmotic pressure of the medium) by absorbing or excreting salts until the osmotic pressure inside and outside the animal is the same; in this way the water-content of the body remains steady; and adaptations for living in 'wetter' or 'drier' places depend on the ability of the cells of the body to carry on their normal processes while bathed in fluids of different osmotic pressures.

But there are some aquatic species whose blood varies little in osmotic pressure despite wide variations in the concentration of the medium in which they live; and there are species that form a graded series between these two extremes. Those species that maintain the water-content of their bodies fairly steady while the blood continues to have a different osmotic pressure from the external medium, are able actively to transport salts across one or other membrane in their bodies. There is no convincing evidence that any aquatic animal transports water actively across a membrane. Water is moved by swallowing or excreting or by osmosis; salts are moved across membranes against gradients in concentration – sometimes quite substantial ones.

The brine shrimp *Artemia salina* can survive in a medium ranging from 10% sea-water up to a saturated brine from which the salts are crystallizing out (Croghan, 1958a). As the concentration of the medium varied through this range the water-content of the animals remained virtually constant, changing only from 86 to 89%. The eel *Anguilla vulgare*, as a normal part of its life, goes from the sea into fresh water and back into the sea again (sec. 4.15). When eels that had been living in fresh water were placed abruptly into sea-water their weight increased at first; during the first 10 hours they increased in weight by 4%, but during the next 40 hours they resumed their normal weight and thenceforth remained in equilibrium with sea-water (Krogh, 1939). The mechanisms whereby these adjustments are made will be discussed in section 5.221.

The pupae of *Ephestia kuhniella* that were reared in oven-dry flour contained 64% of water compared with 68% in pupae that had been reared in flour containing 10% of water (Fraenkel and Blewett, 1944). The kangaroo rat *Dipodomys merriami*, living in the desert in Arizona on air-dry barley, contained 66% of water compared with 67% when it had a moist diet. For comparison an ordinary white rat that was allowed to drink as much water as it would contained 69% of water (K. Schmidt-Nielsen and B. Schmidt-Nielsen, 1952). When *Dipodomys* were desiccated by being offered nothing but dry soy bean (40% nitrogen) for food and nothing to drink they lost about 34% of their original weight before dying, but at death they still contained 67% water. When *Tribolium confusum* were starved in dry air (0% relative humidity) for 7 days they lost weight steadily, losing about 30% of their original weight in 7 days; but the proportion of water in their bodies remained constant at about 56% (Roth and Willis, 1951). Few terrestrial animals are as hardy as *Ephestia, Tribolium* and *Dipodomys* but most species, delicate and hardy ones alike, need to maintain a relatively high level of water in their bodies if they are to remain alive.

Important exceptions to this rule relate chiefly to animals in a condition of diapause, anabiosis or some other form of dormancy. For example, the fully grown larva of the moth *Diacrisia*, which overwinters as a diapausing caterpillar, changed in weight from 600 to 200 mg. in the autumn. Water accounted for most of the weight that was lost. In the spring it absorbed water and nearly regained its original weight before pupating (Payne, 1927). Similar or even greater reductions in water-content may occur in rotifers, tardigrades and nematodes as they become dormant. Some of these animals are extremely drought and cold-hardy in this condition and may remain alive but dormant for years or decades (Preyer, 1891). Certain earthworms are reported to have survived brief periods of desiccation down to 70–80% of their original water (Schmidt, 1918; Hall, 1922); and Thorsen and Svihla (1943) claimed that certain frogs survived brief periods of desiccation down to 38–60% of their original water. The camel is remarkable among mammals because it can sustain its normal activities while losing water equal to 30% of its weight and it may be desiccated and rehydrated repeatedly as a normal part of its life (Schmidt-Nielsen *et al.*, 1956).

These are important exceptions, but the general rule is that most terrestrial animals cannot survive severe desiccation of their tissues. And most adaptations for life in dry places comprise either patterns of behaviour that allow the animal to avoid the worst extremes of dryness, or physiological mechanisms that reduce the amount of water lost by evaporation, respiration and excretion.

5.21 *Patterns of behaviour in relation to moisture*

Moisture of the air in the environment of a terrestrial animal may be conveniently measured either as relative humidity or evaporation (Fig. 5.07). Wellington (1949b) postulated that the behaviour of the caterpillars of *Choristoneura* was more closely related to evaporation than to relative humidity, and that the response to moisture would outweigh the response to temperature. Accordingly he designed an apparatus in which he could control temperature and humidity at the same time, producing a gradient in either or both, and he designed a small evaporimeter that measured the rate of evaporation directly. He placed four batches of larvae of *Choristoneura* in four different sorts of gradient which are described in the legend to Figure 5.11. In each experiment most of the larvae congregated in a zone where the evaporation was between 0·12–0·14 cu. mm. per minute despite big variations in temperature and absolute humidity between experiments. To some extent the response may have depended on the condition of the larvae (although this was not brought out in this experiment) because, on another occasion, Wellington found that similar larvae, that had been starved in dry air for one hour, tended to congregate in a drier zone than larvae that had not been starved.

Roth and Willis (1951) showed that the behaviour of the flour beetles *Tribolium confusum* and *T. castaneum* in a gradient of moisture depended on the condition of the beetles. The insects were placed in an 'olfactometer' which allowed them to move either into a stream of moist air (nearly saturated with water vapour) or a stream of dry air (near 0% relative humidity). Some beetles were taken from the flour containing 10–12% water where they had been living and tested immediately; others were starved in air at about 0% relative humidity for periods ranging from 1 to 6 days. The results are shown in Figure 5.12: 6 days' starvation reversed the normal tendency to congregate in dry air; both species were alike in this but *T. confusum* was more 'hardy' than *T. castaneum* because a longer period of starvation was required to produce the same result (exp. 12.11).

Roth and Willis carried these experiments one step further. Beetles (females of *T. castaneum*) that had been starved in dry air for 4 days (by which time only 7% of them were in the condition to choose dry air) were then starved for a further 60–70 hours in moist air (with a relative humidity in excess of 90%) during which time they continued to lose a little weight (from 73 to 72% of the original weight) but the proportion of water in the body increased slightly, from 60 to 62%. When these beetles were tested in the olfactometer 42% chose the dry air, i.e. 60–70 hours' starvation in moist air had allowed 35% of the beetles to regain their preference for dry air. Another lot of beetles (males of *T. confusum*) were starved in dry air for 6 days; at the end of this time 87% preferred the moist air. They were then placed in flour

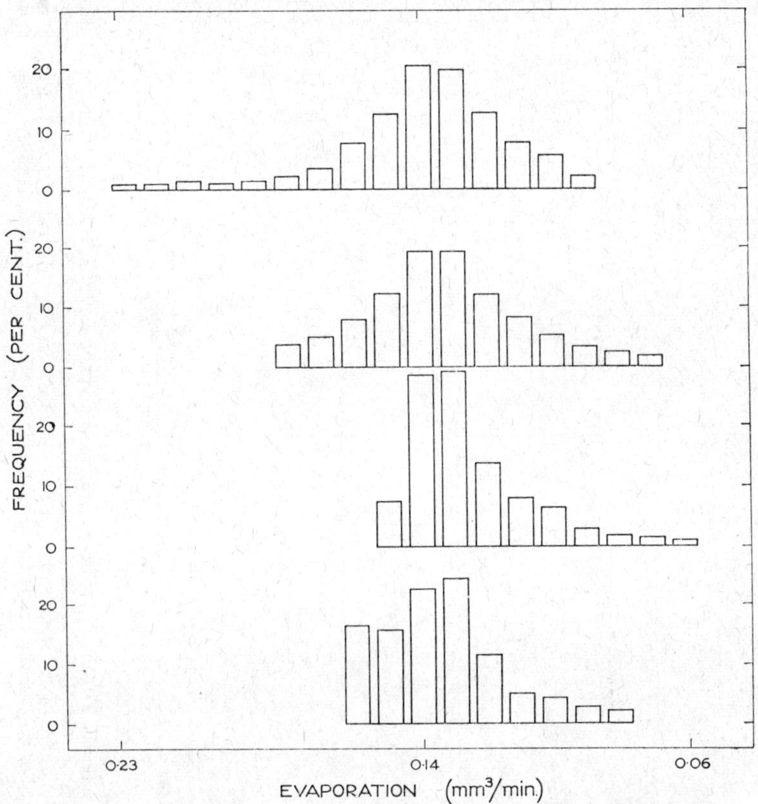

Fig. 5.11. Distribution of second-instar larvae of the spruce budworm *Choristoneura fumiferana* in response to evaporation in four sorts of gradient: A (*top*), in a gradient of temperature in dry air. The isotherm for 36°C lay above the zone where the evaporation was 0·23 mm.³/minute. B, in a gradient of temperature in moist air. The isotherm for 36°C lay above the zone where the evaporation was 0·18 mm.³/minute. C, in a gradient of temperature in a very moist atmosphere. The isotherm for 36°C lay above the zone where the evaporation was 0·15 mm.³/minute. D (*bottom*), in a gradient of evaporation at a constant temperature of 20·6°C. The maximal rate of evaporation did not exceed 0·16 mm.³/minute. (After Wellington, 1949b)

that had been oven-dried to a moisture-content of 0·13% and kept for 66 hours in this flour in a desiccator containing anhydrous calcium chloride. After this treatment 67% of the beetles preferred the dry air.

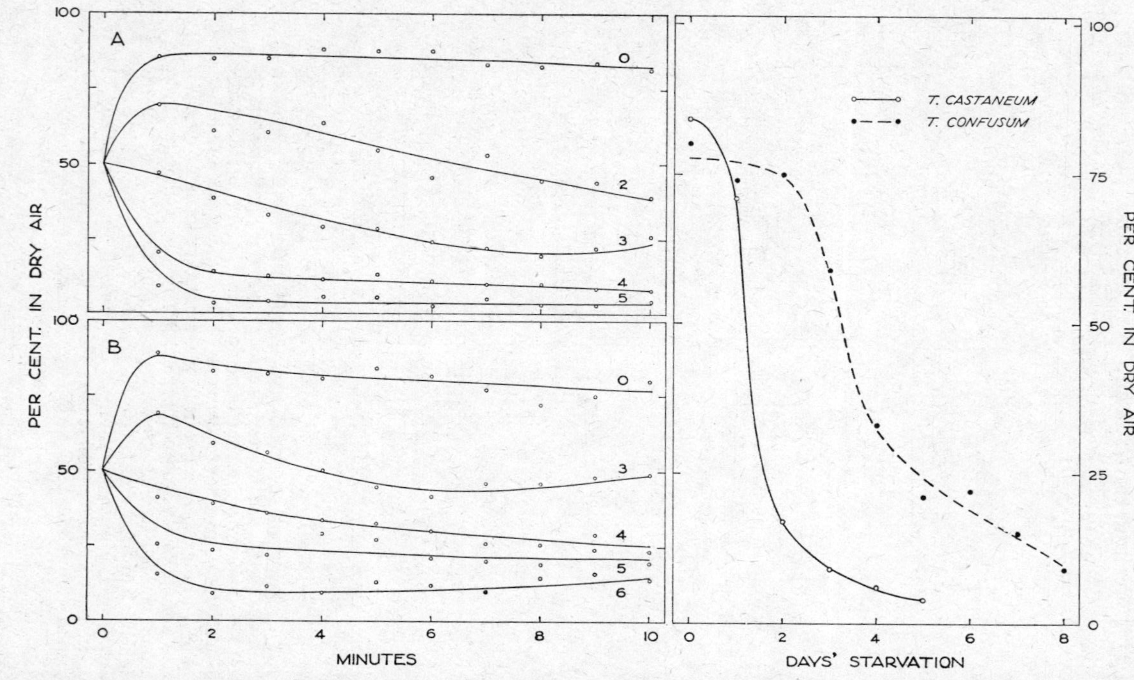

Fig. 5.12. The preference of adult beetles, *Tribolium*, for moist or dry air and the way that this may change in beetles that have been starved and desiccated. A, *T. castaneum*. B, *T. confusum*. The ordinates show the per cent of beetles on the dry side of the olfactometer. The figures against the curves show the number of days' starvation before the test began. The curves on the right show the period of starvation required to reverse the beetles' normal preference for dry air. (After Andrewartha and Birch, 1954)

The opportunity to eat, even though the food was oven-dry flour, had so altered the condition of the beetles that 54% of them had regained their normal preference for dry air (see *Ephestia*, sec. 5.2).

The moisture in the environment of an animal that is living in soil is best measured in terms of the force required to 'pull' water out of the soil. One way to measure this force is to place some damp soil on a porous plate from which is suspended a long tube, open at the bottom, containing water. The water in the tube must make good contact with the water in the soil so that the water in the tube is held against gravity by the surface-forces holding the water in the soil. Soil physicists use the logarithm of the height of this column of water as a measure of the moisture-content of the soil; it is sometimes called the pF of the soil but there is a modern trend to drop this name because the analogy with the pH scale has not proved very helpful (Richards, 1949). It is more useful to think of the height of this column of water as representing the 'negative pressure' of the water in the soil. Maelzer's (1958) experiments with the scarabeid *Aphodius howitti* illustrate the usefulness of this way of expressing the water content of soil.

The life-cycle from egg to adult of *Aphodius* occupies one year. The whole life-cycle is spent in the soil except that the adults come to the surface to disperse before burrowing into the soil again to lay their eggs. The pupae transform to adults during December which is mid-summer and approaching the height of the dry season in South Australia. They lie at the foot of burrows, about 9 inches below the surface of the soil, until they are stimulated by an increase in moisture to emerge and disperse. Usually a thunderstorm in January or February may bring enough rain to stimulate the beetles. In seeking places to burrow into the soil again and lay their eggs the beetles congregate quite strongly in particular areas. Maelzer postulated that the strong preference shown by the beetles for particular areas was associated with the moisture-content of the soil and he did the following experiment.

A number of small boxes, open at the top, were filled with two sorts of soil, a sand and a clay-loam. The water-content of the sand varied from 2½ to 12% and that of the clay-loam from 17 to 37%. The boxes were arranged at random in a large cage and a number of female *Aphodius* were placed in the cage. Subsequently the numbers of beetles that had burrowed into each box were recorded. Table 5.04 shows that for each soil there was an optimum water-content which was not related to the amount of water in the soil but to the force with which the water was held in the soil: in the sand 79% of the beetles were in the range pF

TABLE 5.04*

The number of female *Aphodius* (as per cent of total) found burrowing in soils of different water contents (expressed as per cent dry weight of soil and as the logarithm of the 'negative pressure' of the water in the soil)

Type of soil	*Moisture-content*		*No. of beetles (% total)*
	%	*Log. neg. press.*	
Sand	11·5	2·3	15
	8·5	2·9	45
	3·5	3·5	34
	2·5	4·0	6
Clay-loam	37·0	2·5	22
	31·0	2·9	38
	25·0	3·2	28
	17·0	3·7	12

* After Maelzer (1958).

2·9–3·5 and in the clay 66% were in the range 2·9–3·2; on this scale the 'permanent wilting coefficient' for plants is about 4·2.

Fraenkel and Gunn (1940) speak of a 'token stimulus' when the behaviour of an animal in response to a particular stimulus does not have survival value in relation to this stimulus but does have survival value in relation to some other component of environment which is indirectly related to the 'token stimulus'. In this sense moisture is a token stimulus to *Aphodius*.

The explanation is as follows. The beetles usually oviposit during a temporary lull in the dry season brought about by a summer thunderstorm. If, as often happens, no more than 25–50 points of rain fall on soil that is 'air-dry' the places most likely to remain moist enough to attract the ovipositing beetles are relatively bare areas carrying the sparse stubble of winter-growing pastures rather than areas where the pastures are dense and include perennials, because a dense dry stubble may arrest 15–25 points of rain without it ever reaching the soil. In the former sort of place the pasture which develops during next winter and spring is not likely to make a dense sward; the feeding of the larvae is likely to make it sparser. Consequently, by the beginning of summer the soil where the mature larvae and subsequently the pupae and adults must lie dormant, waiting for the time when the adults may disperse, is likely to be moister than that under a dense sward of pasture, especially if it includes perennials which continue to draw on the water in the soil late into the dry season. Thus, although these insects are not very hardy they are fairly secure because the behaviour-pattern of the adult (response to a 'token stimulus') ensures that the larvae will be living in places that are relatively moist at the time when the hazards from dryness are greatest.

The kangaroo rat *Dipodomys merriami*, which lives in the desert in Arizona, also has a pattern of behaviour that protects it from exposure to the most extreme hazards of dryness in the desert. B. Schmidt-Nielsen and K. Schmidt-Nielsen (1951) showed that *Dipodomys* could maintain the water-content of its tissues in air at 24°C and 10–20% relative humidity with nothing to drink and no food other than air-dry barley (sec. 5.2).

During early summer when the weather had not yet reached its most extreme the temperature and humidity near the ground during the day varied between 20°–45°C and 5–15% relative humidity. During the night the readings ranged between 15°–25°C and 15–40% relative humidity. The kangaroo rat could not live for long if it were exposed to these daytime temperatures and humidities; but it spends the day curled up in a burrow that is about one foot beneath the surface of the ground, and emerges only at night to forage. B. Schmidt-Nielsen and K. Schmidt-Nielsen (1950) measured the temperature and humidity in the burrow during the day while it was occupied by the rat. They tied a very small thermohygrograph to the tail of a rat and let it go at the mouth of its burrow. Twelve hours later they excavated the burrow and retrieved the instrument. The maximum temperature varied from 25° to 30°C and the relative humidity from 30 to 50%. The absolute humidity varied from 8 to 15 mg. per litre inside the burrow and from 1 to 5 mg. per litre outside the burrow.

B. Schmidt-Nielsen and K. Schmidt-Nielsen (1949) calculated the amount of water that *Dipodomys* saved by staying in its burrow during the day:

> We can assume that *Dipodomys*, like other mammals, expires air that is saturated with moisture slightly below body temperature. Normal body temperature for *Dipodomys merriami* is 36°–37°C and the following calculations are made under the assumption that the expired air has a moisture content corresponding to saturation at 33°C (Dill, 1938, p. 26). Air saturated with water vapour at 33°C contains 35·3 mg. of water per litre of air. A *Dipodomys* that is breathing

completely dry air will, therefore, lose 35 mg. of water per litre of air expired. If the animal breathes desert air with a moisture-content, as we measured it during the daytime, of about 2 mg. of water per litre of air, it will lose approximately 33 mg. of water per litre of air expired. If, however, the animal stays in its burrow, the air it inspires will contain about 10 mg. water per litre, and the animal will lose only about 25 mg. water per litre of expired air, or 24% less than in the first case.

It is obvious that a high moisture-content of the inspired air results in a lower evaporative loss from the lungs, and that the absolute humidity and not the relative, is the determining factor. In this way a relatively dry and warm burrow may be more advantageous than a cold and relatively humid burrow, because the latter may have a lower absolute humidity.

5.22 *Physiological mechanisms for conserving water*

5.221 'Water-balance' in aquatic animals

Krogh (1939) discussed this subject thoroughly. I have taken the following brief summary largely from his book. It is also discussed by Ramsay (1952, Chap. 4).

Most marine invertebrates have the fluids in their bodies isotonic with the water outside. Some species that live in estuaries and in other places where the salinity of the water may vary widely also do this. They can live in such places by virtue of adaptations that allow them to carry on their life-processes when their cells are bathed in fluids with a wide range of osmotic pressures. Although the osmotic pressure of the blood may vary widely the concentration of particular ions, e.g. potassium, may remain more constant. So long as the concentration of salts in their blood remains within the range that they can tolerate the proportion of water in their bodies remains fairly constant. At extreme concentrations they die, but it seems more appropriate to think of them dying from imbalance of salts rather than from excess or lack of water.

All marine vertebrates and some invertebrates maintain their blood hypotonic to the water by excreting salts. The invertebrates in this group form a graded series from those in which the osmotic pressure of the blood is only slightly different from the water and usually responds to changes in the salinity of the water, to those in which the blood is quite markedly hypotonic to the water and may be largely independent of changes in the salinity of the water. In teleost fish salts are excreted chiefly by the gills. Water is lost with the urine and by osmosis mostly through the gills. It is replaced by drinking and with the food.

Nearly all the animals that live in fresh water absorb more or less water by osmosis and also gain some with their food. The whole of this has to be got rid of by excretion. The excretion of water tends to drain salts out of the body. Salts that are lost in this way may be replaced with the food. Certain Crustacea, e.g. *Daphnia*, can live for days in distilled water without food. This implies a remarkable capacity for extracting salts from the urine before it is excreted. Certain Crustacea, including *Daphnia*, can also absorb salts from extremely dilute concentrations in the surrounding water. In the mosquito larva the anal papillae are permeable to water and they also absorb salts from the water; the rest of the cuticle is highly impermeable (Wigglesworth, 1933; Koch, 1938). Krogh (1937) showed that a number of fresh water fish could absorb salts from the water. It seems that most of the animals that live in fresh water follow the same pattern. Many develop a cuticle that restricts the movement of water into the body by osmosis; and they maintain the water-content of their bodies and the osmotic pressure of the fluids that bathe their tissues fairly constant, by excreting the water that is gained by osmosis and with the food, by retaining the salts taken in with the food, and by absorbing salts from the water in which they are living. The adult insects that live in fresh water, e.g. *Hydrophilus*, are exceptions to this rule. They breathe air through the tracheal system and have the entire body covered by an impermeable cuticle that virtually prevents the entry of any water by osmosis.

The eel, the salmon and other teleost fish are remarkable because they move from the sea to fresh water and back again as a regular part of their life-cycle. While living in fresh water the eel drinks no water and secretes a copious dilute urine which contains scarcely any salts. In sea-water the eel does not eat but it replaces the water that is lost by osmosis by drinking sea-water; and it maintains a relatively constant osmotic pressure internally by excreting salts like other marine fish.

5.222 Conservation of water in terrestrial insects and ticks

Insects that live in dry places may retain their nitrogenous waste-products as insoluble urates or they may excrete them with the faeces which are dried out by special rectal glands (Wigglesworth, 1932). So virtually the only water that is lost by insects that live in dry places is that which evaporates through the cuticle or is lost from the spiracles during respiration.

Insects are remarkable because they lose so little water through the cuticle despite their small size. For example, the egg of the grasshopper *Austroicetes* is about 2 mm. long by 1 mm. diameter. It weighs about 4·9 mg. of which 3·3 mg. is water. While it remains in diapause it can survive the loss of 2·5 mg. but will die if it loses much more than this. In nature, the eggs are laid in hard bare soil about half an inch below the surface; they remain dormant in the soil during the long hot, dry summer, hatching in the following spring. I have estimated that a standard Australian evaporimeter (which is a tank 3 feet in diameter sunk into the ground until its surface is level with the ground) would evaporate 3 feet of water while 50% of the eggs were losing enough water to kill them. That is, the average egg would have lost 2·5 mg. while the evaporimeter was evaporating 3 feet.

Matthee (1951) found that the extreme drought-hardiness of the egg of the locust *Locustana pardalina* is partly explained by the relative impermeability of the cuticle that covers the entire egg except for a small permeable 'hydropyle' at the posterior pole, through which water is taken up when the soil is moistened by rain. The impermeability

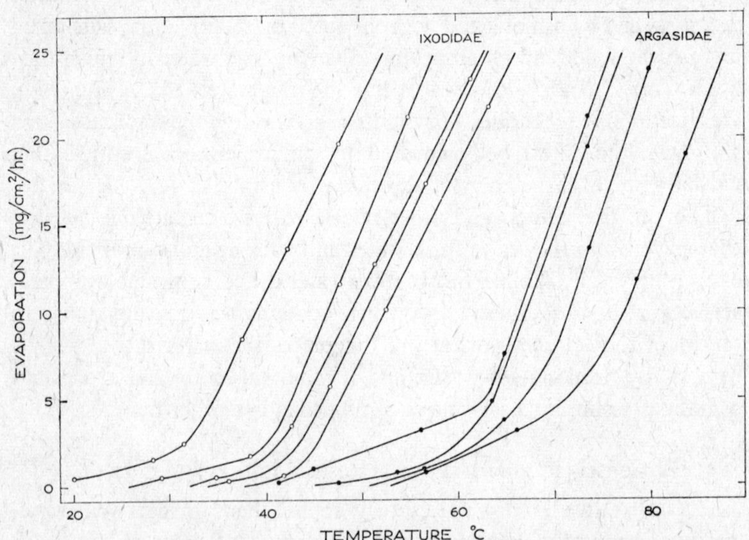

Fig. 5.13. The rate of evaporation of water from dead engorged ticks at different temperatures. The curves refer, from left to right, to the following species. *Ixodes ricinus*, *I. hexagonus*, *Amblyomma americanum*, *Dermacentor andersoni*, *Hyalomma savignyi* (Ixodidae), *Ornithodorus moubata*, *Argas persicus*, *Ornithodorus acinus*, *O. savignyi* (Argasidae). (After Lees, 1947)

of the cuticle is largely due to a layer of wax incorporated deep in the cuticle. The impermeability of the cuticle of insects and ticks is also due to a layer of wax which usually forms the outermost layer but one in the epicuticle; it may be covered and protected by a thin layer of 'cement' (Wigglesworth, 1948).

Lees (1947) measured the rate at which water evaporated through the cuticle of several species of ticks. The ticks were dead and the openings of the respiratory and alimentary systems had been blocked with wax; so it is likely that most of the water was lost through the cuticle. Figure 5.13 shows that at moderate temperatures the cuticles of all the ticks transmitted rather less than 1 mg. of water per square cm. per hour. But for each species there was a narrow range of temperature in which the rate of evaporation increased enormously. In general this critical zone occurred at higher temperatures in the Argasidae which normally live in much drier places than the Ixodidae. Lees showed that these differences were associated with waxes of different melting points.

But the water-balance of an insect or a tick that is starving in dry air is not to be explained merely in terms of the physical permeability of the cuticle. Insects and ticks can use the energy of respiration to prevent evaporation from their bodies or to absorb water-vapour from air against substantial gradients of pressure. Lees (1946) exposed the tick *Ixodes* to air at 25°C and 50% relative humidity until they had lost about 10–20% of their original weight. He then placed them at 95% relative humidity and found that they rapidly regained their original weight by absorbing water-vapour from the air. By a series of similar experiments he found that the critical humidity for *Ixodes* was about 92%. The ticks could be starved in air at 92% relative humidity without losing water; alternatively ticks that had been desiccated could absorb water-vapour from air in which the relative humidity exceeded 92%.

Browning (1954) exposed the tick *Ornithodorus moubata* alternatively to 0% and 95% relative humidity at 25°C as indicated in Figure 5.14. The ticks lost weight at 0% and regained it rapidly at 95% humidity; they continued to do this for 60 days without any apparent change in the maximal weight achieved during exposure to moist air. After 60 days the pattern was the same but the maximal weight fell off slowly until after 140 days they died. The first cause of death was almost certainly exhaustion of 'fuel' rather than shortage of water. When CO_2 was added to air at 0% relative humidity, the ticks lost weight no more rapidly than in air until the concentration of CO_2 reached 30%, which is the concentration that is required for anaesthesia. Similarly in

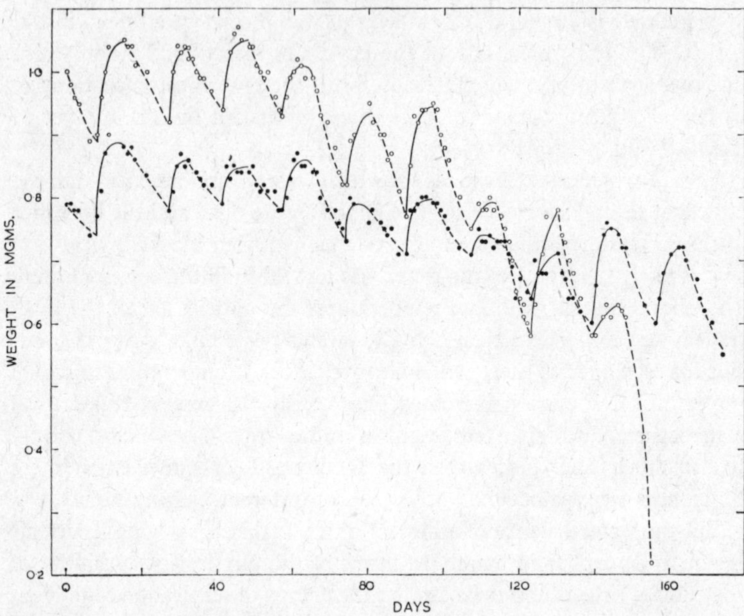

Fig. 5.14. Changes in weight of 2 second-instar nymphs of *Ornithodorus moubata* that were exposed alternately to 95% relative humidity (complete lines) and 5% relative humidity (broken lines). (After Browning, 1954)

mixtures of air and nitrogen the ticks retained their normal ability to resist desiccation until the concentration of nitrogen approached 98%, suggesting that the failure of the ticks was caused by shortage of oxygen. Earlier experiments by Lees (1947) had suggested that the epidermis was the tissue chiefly responsible for the ticks' ability to absorb water-vapour against a gradient of pressure.

Fraenkel and Blewett (1944) reared larvae of the moth *Ephestia* in three lots of flour with water-contents 1·1%, 6·6% and 14·4%. The three lots of flour were in equilibrium with air having relative humidities 0%, 20% and 70%. Table 5.05 shows that the pupae reared in dry flour contained nearly the same proportion of water as those reared in moist flour but those in the dry flour digested more food per mg. of tissue produced. At least some of the water in the pupae that had developed in the moist flour may have come from the food but virtually all the water in the pupae in the dry flour must have been manufactured in the body by oxidizing food. This means that the larvae were able to respire carbon dioxide while retaining some of the water of respiration.

TABLE 5.05*

Relationship between moisture-content of food and amounts of food digested to produce the same weight of dry matter in the tissues of *Ephestia kühniella* reared in flour at 25°C

Moisture in food (%)	*Relative humidity* (%)	*Wet weight of pupa* (*mg.*)	*Water-content of pupa* (%)	*Ratio dry wt. food to dry wt. pupa*
14·4	70	25·5	68	6·3
6·6	20	18·7	66	9·0
1·1	0	15·8	64	12·7

* After Fraenkel and Blewett (1944).

I do not know what proportion was retained because Fraenkel and Blewett did not record the humidity of the air that was respired.

Nothing is known of the mechanism whereby metabolic water is retained in the body. Nor do we know whether it is the same mechanism that allows a fasting insect to resist the evaporation of water into dry air and absorb water-vapour from moist air. It is not even certain that the last two processes are done by the same mechanism.

5.223 Conservation of water in terrestrial mammals

The kangaroo rat *Dipodomys merriami* is the most drought-hardy mammal that is known. In nature, it lives in the desert eating only dry seeds and not drinking any water. In the laboratory, when it was offered no other food than one that was high in nitrogen, as soy bean, it drank; when no other water was available it drank sea-water; it lived for weeks, and even gained weight, on a diet of soy bean and sea-water.

The physiology of *Dipodomys*, especially with respect to the conservation of water, has been studied by K. Schmidt-Nielsen and B. Schmidt-Nielsen (1952, and other papers). They first did experiments which demonstrated that (*a*) *Dipodomys* did not store excretory wastes as harmless insoluble solids in its tissues. (*b*) It did not carry in its body a reserve of water that could be drawn on in an emergency. (*c*) It did not survive an unusual reduction in the water-content of its body (sec. 5.2). In short, *Dipodomys* used water in just the same way as any other mammal (except that it did not sweat), but it was very economical.

Table 5.06 compares the water-economy of *Dipodomys* with that of *Rattus* and man. The difference between *Dipodomys* and *Rattus* in the amount of water lost by evaporation seems too big to be accounted for by the little sweating done by *Rattus*. There was no evidence from breathing of *Dipodomys* or from the oxygen-dissociation curve of its blood to suggest that it used oxygen any more efficiently than *Rattus*.

TABLE 5.06

Water-economy of *Dipodomys* compared with that of *Rattus* and man

	Dipodomys	Rattus	*Man*
Evaporation from all sources (mg. H_2O per ml. O_2)	0·54*	0·94†	‡
Concentration of urine (urea)	3·8M§	2·5M	1·0M
Total electrolytes	1·5N‖	0·6N	0·37N
Water-content of faeces	45%	68%	
Water lost with faeces per 100 gm. barley	2·5 gm.	13·5 gm.	

* None lost by sweat.
† A little lost by sweat.
‡ May lose 10 times as much from sweat as from all other sources combined.
§ Equivalent to 23% urea.
‖ Equivalent to 8·5% NaCl.

But it was suggested that the shape of the nose in *Dipodomys* may be an adaptation for conserving water. The nose is long and thin and poorly supplied with blood. When the temperature of the air is below 37°C (the body temperature of *Dipodomys*) the temperature of the nasal passages may be lower than the temperature of the lungs which may allow some of the water of respiration to condense and be re-absorbed in the nose.

The total intake of water comprises that extracted from the food, that manufactured by oxidizing the food, and that inhaled during breathing. The total output of water comprises that evaporated from skin and lungs, that excreted with the urine and that excreted with the faeces. Because all these quantities were known, or could be calculated, K. Schmidt-Nielsen and B. Schmidt-Nielsen (1952) were able to draw up a complete 'balance sheet' of the water-economy of *Dipodomys*, which I have reproduced in Figure 5.15. The lines for intake and output intersect at 10% relative humidity which is the lowest humidity at which *Dipodomys* can live at 25°C when it has only dry barley for food and no water to drink.

In deserts in other parts of the world there are other species of small rodents that seem to live the same sort of life as *Dipodomys*. Although they have not been studied I expect that they may turn out to be drought-hardy like *Dipodomys*.

The camel is well known for its ability to live and work in the desert (B. Schmidt-Nielsen *et al.*, 1956). The camel reduces very greatly the amount of water used for evaporative cooling by having thick, woolly fur that insulates it from the radiant heat of the sun, and by tolerating large variations in the temperature of its body. In the

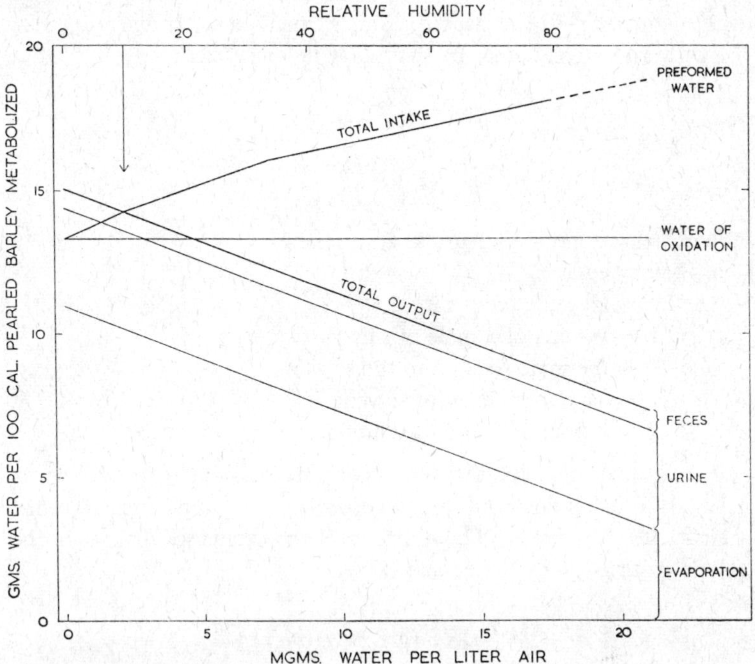

Fig. 5.15. 'Balance sheet' of the water-economy of *Dipodomys*. For explanation see text. (After Schmidt-Nielsen and Schmidt-Nielsen, 1951)

morning the temperature of the body may be as low as 34°C but, during the heat of the day, the camel sweats scarcely at all until its temperature exceeds about 40°C. The camel also tolerates severe desiccation compared with other mammals; a man will die from 'explosive heat shock' when he has lost water equivalent to 12% of his original weight; but a camel may lose up to 30% and still continue to function normally.

It is possible to compare the performances of the animals that have been mentioned above by reducing their performances to a common denominator. For example, the flea *Xenopsylla* is able to absorb water from air at 50% relative humidity and secrete it into its tissue fluids that have an osmotic pressure of about 8 atmospheres. A solution that was in equilibrium with air at 50% relative humidity would have an osmotic pressure of 910 atmospheres. The tissue-fluids of *Xenopsylla* contain dissolved salts equivalent to about 1% NaCl, i.e. about 0·2N. Now relative humidity, osmotic pressure and concentration of dissolved salts are merely different scales in which the *activity* of the water in the environment may be measured. Measurements taken in one scale may be converted to any other by the following equations:

$$\pi = -\frac{RT}{v}\log_e(1 - X)$$

$$= RT \log \frac{P_0}{P}$$

Where π = osmotic pressure in atmospheres
R = 0·08206
T = absolute temperature
v = volume of 1 mole of solvent (water)
X = the mole fraction of the solute
P_0 = vapour pressure of solvent
P = vapour pressure of solution

Lees (1947) suggested that one way that *Ixodes* might transfer water from air to its tissues would be to absorb it in fine capillaries and then transfer it to its tissues. The diameter of the capillaries that would be required are given by the equation:

$$r = \frac{2\gamma M}{DRT(\log_e p_1 - \log_e p)}$$

where r = diameter of capillary
γ = surface tension of water
M = molecular weight of water
D = density of water
R = the gas constant
T = absolute temperature
p_1 = vapour pressure at curved surface
p = vapour pressure at plane surface

The figures in Table 5.07 have been calculated from these equations. They serve to compare these animals with respect to the gradient of pressure against which they can cause water to move into their bodies. In terms of pressures *Xenopsylla* is able to move water from a solution with osmotic pressure of 910 atmospheres into a solution with osmotic pressure of 8 atmospheres. The flea is doing work against a gradient in pressure of 902 atmospheres. Table 5.07 compares the 'performance' of *Xenopsylla* with that of a number of other animals that can live in dry places. This comparison does not take account of the work that is done

TABLE 5.07

The dryness of the environment measured in terms of the 'activity' of water in the environment

Species	*Empirical information*	*Estimated 'dryness' of environment*			
		Osmotic press. atmos.	*NaCl equiv/liter*	*Rel. hum. (%)*	*Diam. capil. (A°)*
Rat flea[1]	Gains wt in air R.H. 50%	910	*	50†	32
Brine shrimp[2]	Maintains body fluids in 30% NaCl	326	*	79	
Camel tick[3]	Gains wt in air R.H. 85%	216	3·3N	85†	136
Kangaroo rat[4]	Concentrates urine urea 3·8M; NaCl 1·5N	205	3·2N	86	
Sheep tick[3]	Gains wt in air R.H. 92%	109	1·9N	92†	260
White rat[4]	Concentrates urine urea 2·5M; NaCl 0·6N	99	1·6N	93	
Man[4]	Concentrates urine urea 1·0M; NaCl 0·37N	44	0·87N	97	
Gull[5]	Excretes salt NaCl 5%	41	0·85N	97	
Eel[6]	Excretes salt NaCl 3·3%	26	0·56N	98	

* Beyond the solubility of NaCl.

† Empirical observation.

[1] Edney (1947); [2] Croghan (1958b); [3] Lees (1947); [4] K. Schmidt-Nielsen and B. Schmidt-Nielsen (1952); [5] Schmidt-Nielsen and Sladen (1958); [6] Krogh (1939).

in maintaining the water-content of the body or the concentration of salts in the blood, but only the gradient in pressure against which the work is done. The larvae of *Ephestia*, living in oven-dry flour and growing up to pupate with 64% of water in their bodies (Table 5.05), undoubtedly do a lot of work but I did not include them in Table 5.07 because I do not know to what extent they dried the air before they expired it. In *Artemia*, as the water in which they were living changed from 10% sea-water to saturated brine, the concentration of sodium in the haemolymph doubled but the concentration of potassium increased by about 10%. To maintain such a gradient in the relative concentration of different ions may require additional work but the difference in relative concentrations does not in itself influence the magnitude of the pressure against which the work has to be done.

5.3 LIGHT

The exciting aspect of light, to the animal ecologist, is the diverse ways in which animals have become adapted to respond to light as a 'token stimulus', especially the ways in which it serves as a 'clock' by which life-cycles are synchronized with each other or with the seasons. Physiological studies, especially of the endocrine organs of mammals, birds and insects, have greatly enhanced our understanding of light as a component of environment just as our understanding of moisture has been helped by physiological studies of osmoregulation and active transport.

5.31 *The influence of light in synchronizing life-cycles with each other*

The cockroach *Periplaneta americana* is nocturnal. If it is living in a place where daylight penetrates it will remain quiet during the day and become active at night. But when a number of *Periplaneta*, that had previously been exposed to the natural diurnal rhythm of light, were kept in continuous darkness they continued to be quiet during the day and active at night for about 4 days. After this their behaviour lost its regular rhythm (Harker, 1956). When another batch was exposed to artificial light during the natural night and kept in darkness during the natural day they were quiet during the night and active during the day; and this rhythm also persisted for several days when they were kept in continuous darkness. By grafting a legless cockroach that had been conditioned for a certain rhythm on to the back of another that had lost its rhythm (by living in continuous light) Harker showed that the rhythm was caused by a hormone originating in the suboesophogeal ganglion and circulating in the blood.

Bateman (1955) showed that a rhythm imposed on one generation of the fruit-fly *Dacus* was transmitted to the pupae of the next generation. He illuminated two batches of adult flies as follows: one batch was kept in the light from 9 a.m. to 5 p.m. and in the dark for the rest of the day; the other batch was illuminated from 9 p.m. to 5 a.m. The larvae and pupae were kept in continuous darkness, and the adults of the next generation emerged in darkness. Most of the emergences occurred during a period of 48 hours but the emergences were not distributed evenly over this period; on the contrary there were two peaks of emergence separated by 24 hours; the striking thing about them was that they occurred in each batch during that time of the day when the adults of the previous generation had been illuminated. This remarkable example of a 'remembered' rhythm is illustrated in Figure 5.16. I

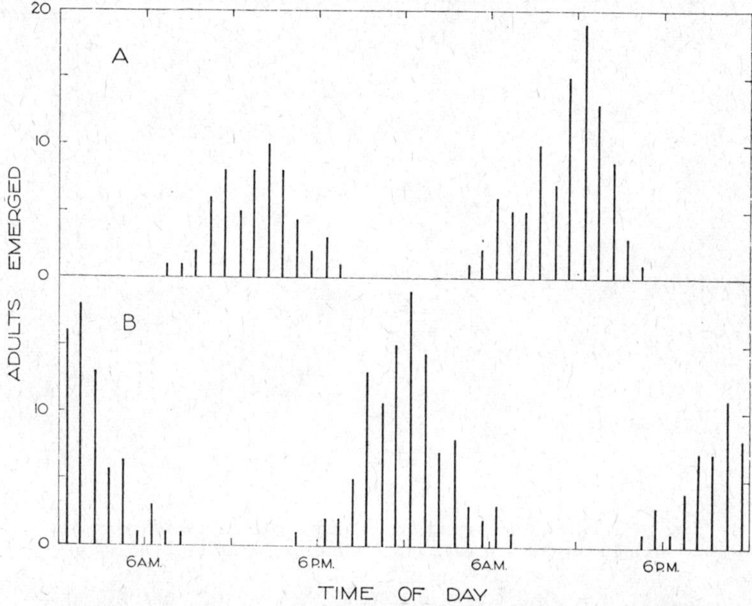

Fig. 5.16. The influence of exposure to light during the adult stage on the 'timing' of the pupal ecdysis in the next generation, in *Dacus tryoni*. A, adults were illuminated from 9.00 a.m. to 5.00 p.m.; B, adults were illuminated from 9.00 p.m. to 5.00 a.m. (After Bateman, 1955)

expect that this rhythm will also be found to be controlled by hormones, but at present we know nothing of the mechanism.

5.32 *The influence of light in synchronizing the life-cycle with the season of the year*

The caterpillars of *Grapholitha molesta* are serious pests of peaches in certain parts of the world. One of the reasons why *Grapholitha* can maintain such high numbers, except when it is sprayed with insecticides or 'controlled' by predators, is that its life-cycle is nicely synchronized with the seasons and with the life-history of its food-plant the peach. During summer there may be several generations and all stages of the life-cycle may be present at once. But during winter when there is no food and when temperature may be low enough to prevent development or even to kill the active stages, the whole population comprises only fully mature caterpillars in a dormant cold-hardy condition which is known as 'diapause' (Dickson, 1949).

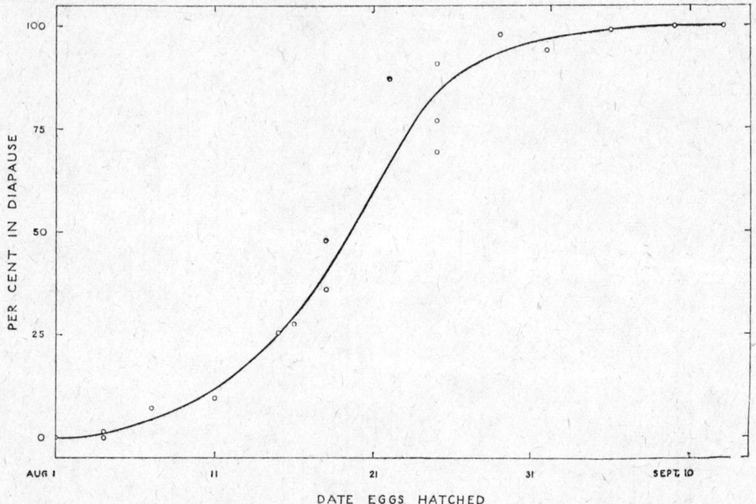

Fig. 5.17. The seasonal incidence of diapause in *Grapholitha molesta* reared out-of-doors at Riverside, California. (After Dickson, 1949)

Diapause is a condition of arrested development: morphologically, the caterpillar is ready to moult to the pupa, but there is a physiological 'block' because the hormone that normally organizes the moult is lacking. The hormone is lacking because a particular group of neurosecretory cells at the base of the brain which normally secrete this hormone have become incompetent to do so. They remain incompetent so long as the caterpillar is kept warm; they regain competence only after a prolonged exposure to low temperature such as they experience during the winter.

In addition to this physiological 'block' the caterpillar in diapause also shows certain other qualities – there is a smaller proportion of water in its body, it is more cold-hardy (secs. 5.12, 5.2), and it develops a particular pattern of behaviour which leads it to seek a well-sheltered place in which to spend the winter. Diapause is found in many species of insects, mites and ticks, and an analogous condition is found in Crustacea, nematodes, rotifers, tardigrades, earthworms, snails and perhaps frogs.

The life-cycle of *Grapholitha* is so nicely synchronized to the seasons because the young caterpillars respond to length of day in a particular way: it seems as if the shortening days of autumn serve as a 'clock' to predict the approach of winter. Dickson (1949) collected young larvae from orchards in southern California, reared them and recorded the proportion entering diapause. Figure 5.17 shows that virtually all those that hatched from eggs after about August 25 entered diapause. At this time of the year there was little more than 12 hours of daylight each day. Dickson then did a series of experiments at 24°C varying the 'length of day' artificially from 0 to 24 hours. The results, shown in Figure 5.18, confirm the observations that were made in the field.

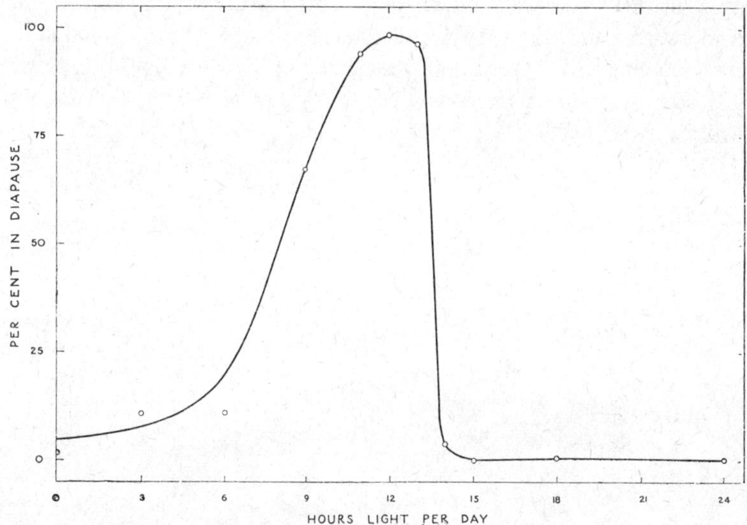

Fig. 5.18. The influence of photoperiod at 24°C on the inception of diapause in *Grapholitha molesta*. (After Dickson, 1949)

Rowan (1938) kept the finch *Junco hyemalis* in outdoor aviaries exposed to the weather. One aviary was artificially lighted by two 50-watt globes which, from October, were switched on at sunset and each day were left on for 5 minutes longer than the day before. By mid-November the testes of the birds in this aviary were increasing in size and by the end of December they were larger than those of normal birds early in spring, despite the fact that the temperature in the aviary had been low – below 0°C at times. The testes of the birds that were not receiving additional light were still quite small. These results have since been confirmed by a number of similar experiments on different species (Marshall, 1942; Bullough, 1951).

Baker and Ranson (1932) found that the field-mouse *Microtus* bred freely when exposed to 15 hours of light each day but virtually ceased breeding at 9 hours. Bissonnette (1935) exposed ferrets to gradually increasing length of day and found that they became sexually mature more rapidly than the controls. Similar results have been reported with hedgehogs, raccoons, lizards, turtles and certain species of fish (Marshall, 1942; Bullough, 1951).

5.4 FURTHER READING

Most of the material in this chapter is discussed rather more fully in Andrewartha and Birch (1954), Chapters 6, 7 and 8. Osmoregulation in aquatic animals is discussed and reviewed by Krogh (1939). Diapause in insects was summarized by Lees (1955); there is also a review by Andrewartha (1952).

CHAPTER VI

Components of Environment: Other Animals – and Pathogens

6.0 INTRODUCTION

In this chapter I have to discuss how an animal's chance to survive and multiply may be influenced by the number of other animals that are associated with it – not only those of its own species but also those of other species; and I also include in this component of environment the diseases that may be caused by bacteria, fungi, viruses, etc. In other books on animal ecology these relationships, or some of them, are usually discussed under the headings of 'intraspecific' and 'interspecific competition'. I do not use these terms because they do not fit into the idea of environment that I defined in Chapter II. Also I think that 'competition' is a misleading concept when it is introduced into the theory of ecology (secs. 7.2, 9.32).

6.1 OTHER ANIMALS OF THE SAME KIND

6.11 *Underpopulation*

The sheep tick *Ixodes ricinus* has been established in northern England for centuries; but there are still some farms that are free from ticks although they seem to be quite well suited to them. It is known that, on occasion, the spread of the ticks from one farm to the next has been held up for decades by no greater barrier than a good sheep fence. It is also known that even after *Ixodes* has become established in a new area it has multiplied slowly, taking many years to reach the level of abundance that is characteristic of populations that have been established for, say, 50 years. Both these problems were studied by Milne (1950, 1951, 1952).

The ticks live for most of their lives in the vegetation near the ground but they do not feed or mate except on a sheep or some other animal. At the appropriate time of the year the ticks emerge from the mat of vegetation and climb to the tip of a protruding grass-stem. If, during the week or so that the tick can stay alive in this position, a sheep brushes by, the tick will cling to the sheep. If a sheep picks up two ticks of opposite sexes they may come together and mate; they are not much good at crawling so their chance of meeting might be quite small if they happen to be picked up at opposite ends of the sheep's body.

Ignoring this last hazard and making certain approximate assumptions, Andrewartha and Birch (1954) calculated the chance that a tick might have of meeting a mate. In a population in which the density was such that on the average each sheep picked up one tick, each one that was picked up by a sheep would have one chance in five of meeting a mate. (The chance is 0·2 and not 0 because by chance some sheep carry 2 or more ticks and some carry none even though the average is one.) As the average number per sheep increased to 10 the average tick's chance of meeting a mate rose to 75%.

These figures do not take account of any other hazards such as the chance that the two sexes will be picked up at different times or on widely separated parts of the sheep. Also Milne estimated that on the average a tick that is just beginning life has about one chance in ten of living to become mature. When all this is taken into account it is easy to appreciate that the occasional tick that is carried through a good sheep fence, by a straying badger or some other animal, would have virtually no chance of founding a colony. If the ticks are to have any chance of becoming established in a new area the original colony must be large; this happens when a whole flock of 'ticky' sheep is moved into a new area.

I do not know of any other species for which this principle has been so nicely worked out. I would think that it is likely to be commonplace among sparse populations of many sorts of animals if only we had a way to observe it. And it seems to me that the success of quarantine measures, which are aimed at preventing the establishment of pests in new areas, must often owe their success to the operation of this principle. On the other hand, the European rabbits which are now such pests in Australia first became established when a few wild rabbits were introduced into Victoria in 1859. Similarly the original introduction of the starling into North America consisted of a very small colony which nevertheless established itself in the new area (sec. 8.2).

Birch and Andrewartha (1941) described the circumstances in which virtually all the grasshoppers *Austroicetes cruciata* that were living in an area of several hundred square miles in South Australia were eaten by birds. The circumstances were unusual because this occurred during a severe drought. The grasshoppers have one generation a year. The eggs are laid in the soil where the birds do not find them. There is a prolonged egg-stage lasting from November to August. The active stages, nymphs and adults, are present from August through November which is the southern spring. We watched this population of grasshoppers for 6 years; from 1935 to 1939 there were dense swarms distributed widespread over the whole of the 'grasshopper belt' (Fig. 1.01). During this period the birds ate freely of grasshoppers during three months of each year but the birds were so few relative to the large number of grasshoppers that only a very small proportion were eaten. The birds did not increase in numbers during this period because there was no comparable supply of food for them during the rest of the year. In August 1940 the grasshoppers hatched from eggs in the same large numbers. But there had been a severe drought; there was scarcely any green grass; and most of the grasshoppers died from starvation. By October the population which had been large was reduced to a series of small remnants, clustered in local situations that still harboured a little grass because they were a little moister than the rest of the country. The birds sought out the grasshoppers and ate them. Because the number of the grasshoppers were few relative to the birds, they were eaten almost to the last one. During the period when the grasshoppers were abundant the chance that any one would be eaten by a bird was small; but during their time of scarcity this risk increased almost to a certainty.

Colonies of the guanay *Phalacrocorax bouganvillii* that are below a certain size do not multiply. It has been suggested that in small colonies the ratio of circumference to area is large and this increases the proportion of chicks that are taken by predators (Hutchinson and Deevey, 1949). Herds of less than about 15 antelopes rarely persist because once the herd has become as small as this it ceases to fight off wolves and coyotes as an organized herd (Leopold, 1933). These are just several examples of the general principle that an animal in a small population may be more likely to be eaten by a 'facultative' predator than one in a large population (sec. 6.213).

Animals in a large population may have an advantage over those in small ones with respect to temperature, moisture and other components of weather. For example, Cragg (1955) found that a newly hatched larva of the sheep blowfly *Lucilia sericata* had little chance of surviving on the skin of a sheep unless the relative humidity remained above 90% for several hours while the maggot penetrated the skin and established a moist place for itself to live. Except in very wet weather scarcely any maggots survived among those that came from egg-masses that had been laid singly on the sheep. But the survival-rate was higher among those that came from aggregations of several egg-masses laid together because the air was more humid in these larger masses. Pearson (1947) measured the rate at which mice *Mus musculus* consumed oxygen at 26°C. At this temperature a substantial proportion of the respiration of a mouse at rest goes towards maintaining the temperature of the body. Pearson compared the oxygen consumed by four mice when they were huddled together and when they were single. The four mice when they were separated used 1·82 times as much oxygen as the same four when they were huddled together. The rate at which oxygen is consumed is an indication of the influence of temperature on the mice. This result showed that the mice experienced a less severe temperature when they were in a group of four than when they were single (sec. 5.0).

Andrewartha and Birch (1954) mentioned several other ways in which the animals in small populations may be at a disadvantage compared with those in dense populations.

In section 9.32 (in the fourth paragraph of the section) I quote the observations of a number of ecologists which confirm the opinion of Darwin: 'Rarity is the attribute of a vast number of species of all classes in all countries.' It follows that the principles of 'underpopulation' which I have discussed in this section may be expected to have a wide application in nature; conversely the principles of 'crowding' which I discuss in the next section might be inferred to have a rather narrower application.

6.12 *Crowding*

I shall leave the discussion of crowding with respect to food until the next chapter (sec. 7.2).

The principles governing crowding relative to living space have largely been worked out in laboratory experiments which I shall mention in section 6.121. In nature relatively few species ever become crowded relative to their 'living space' or to their food. Perhaps the most notable exceptions to this rule are to be found among sessile marine animals, for example barnacles, where crowding may be usual and stunting from lack of space and/or food may often be observed (McDougall, 1943).

In section 6.122 I shall discuss the absence of crowding in the populations of certain vertebrates which manifest 'territorial behaviour'. The fluctuations that have frequently been observed in laboratory populations of *Tribolium*, *Calandra* and other small invertebrates when the experiment has been expressly designed to keep all components of environment constant have not yet been adequately explained. In section 6.123 I shall discuss the possibility that these fluctuations may be analogous with the 'crashes' that occur in natural populations of small mammals.

6.121 Laboratory experiments in 'crowding' relative to 'living room' Park (1948, 1954, and personal communication) kept a number of populations of *Tribolium confusum* and *T. castaneum* in glass vials containing 8 gm. of flour for periods up to 2,100 days (about 70 generations) at 29·5°C. The relative humidity was kept constant at between 60–70%. The flour was sieved out and replaced with fresh every 30 days. A substantial proportion of the food remained at the end of each period. Yet the beetles, having multiplied to about 55 per gm. by the end of 5 generations, never exceeded this number during the rest of the experiment (Fig. 6.01).

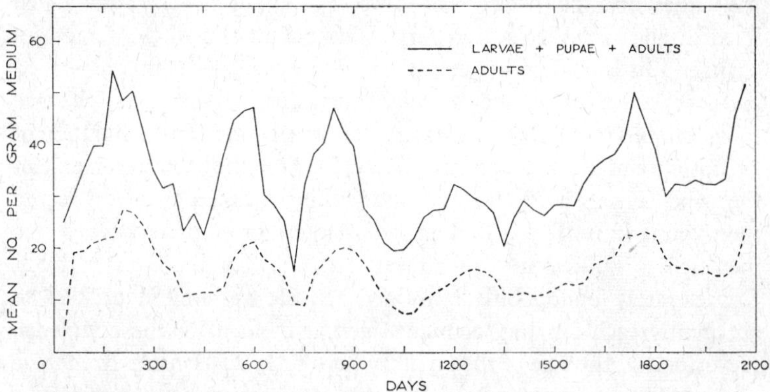

Fig. 6.01. The density of a population of *Tribolium castaneum* living in 8 gm. of flour-yeast medium at 29·5°C and 65% relative humidity. The curves are the means of 20 replicates. (After Andrewartha and Birch, 1954 – data from Park, personal communication)

Because the beetles lived entirely in the flour their 'living room' was restricted to the volume occupied by 8 gm. of flour. Because they never used up all their food between changes it seems likely that their failure to multiply beyond the densities shown in Figure 6.01 was due to crowding relative to living room rather than to food. Park found two ways in which a beetle's chance to survive and multiply may be reduced when it is living in a crowd. (*a*) The larvae and adults of *Tribolium* 'condition' the flour in which they have been living partly by reducing its nutritive value and partly by excreting toxic substances into it;

Tribolium do not live so long or lay so many eggs in 'conditioned' flour as in fresh. (*b*) The adults and larvae eat eggs and pupae; and the chance that an egg or a pupae will be eaten increases as the number of larvae or adults increase.

Similar experiments have been done with *Paramoecium*, *Daphnia*, *Calandra*, *Drosophila* and other species. The results of a number of them were summarized by Allee *et al.* (1949) and Andrewartha and Birch (1954).

6.122 'Territorial' behaviour in vertebrates

The Australian magpie *Gymnorhina dorsalis* is a black and white bird about the size of a pigeon. It is one of the commonest birds in Australia. Carrick (personal communication) studied the ecology of a population of magpies in several thousand acres of farmland near Canberra. Carrick found that the birds in this area comprised two groups which he called the 'tribes' and the 'flock'. A tribe was a group of 2 to 12 birds which lived in and defended a 'territory'. The boundaries of the territories changed a little during the years as a result of vicissitudes that happened either to the tribe itself or to its neighbours. But, in general, a tribe, once it had established itself in a territory, remained there for the duration of its existence as a tribe. The territory sufficed for all the requirements of the tribe. They found all their food within its boundaries and reared their young in it. The minimum requirement for a territory was a tree in which to build a nest, but most territories contained many trees. Carrick rarely found a bird from the tribes that was undernourished, so apparently each territory contained abundant food for the occupants. Nevertheless, the size of the territory bore little relationship to the size of the tribe or to anything else that could be measured. It seemed as if the size of the territory depended chiefly on the innate qualities of the birds, especially their capacity to 'dominate' their neighbours. An established tribe never accepted an immigrant except occasionally to replace a female that had died, nor allowed any of its own offspring to remain in the territory after they had matured. Probably the tribes never lost any members by emigration, but this was more difficult to measure. Individuals died but the survivors continued in the territory until such time as they could no longer defend it; then the whole group would disappear. Sometimes the death of a single bird would lead to the summary eviction of the survivors perhaps because the one that died had been the leader. Occasionally Carrick found birds from a tribe that had been 'evicted' living as individuals in the 'flock'.

The other group of birds, which Carrick called the 'flock', comprised all those that were not members of a tribe with an established territory. None of these birds bred. Apparently the stimulus of the changing seasons was in itself not sufficient to bring the birds to full sexual maturity; they needed, in addition, some stimulus that came from being a member of a tribe in an established territory. The numbers of the flock were swelled each year, towards the end of winter, by the influx of the young birds that had been reared by the tribes during the previous spring. The flock was also increased from time to time by the remnants of tribes that had abandoned, or been driven out of their territories. From time to time a 'tribe' would originate in the flock and establish itself in a territory.

The birds in the flock were more likely to die from parasites, diseases and other hazards than those in the tribes and their supply of food was less certain, but the flock persisted because of the regular influx of young birds from the territories each year. On the other hand, the size of the flock had very little influence on the size of the breeding population which seemed to be determined more by the innate 'territorial' behaviour of the birds than by anything that could be measured. The mechanism seems to provide a wonderful 'buffer' against all components of environment.

Kluijver (1951) described 'territorial' behaviour in a population of great tits *Parus major* living in a mixed deciduous wood near Arnhem, Netherlands. At the beginning of autumn the population comprised old birds that had nested there during summer and their progeny. At this time of the year the birds formed pairs and each pair defended a territory. The 'fighting' was not overt. A bird would fly towards its opponent and turn upwards in such a way as to display a patch of colour on its throat. In the end the 'weaker' bird would give up and fly away. Usually a substantial proportion of the population would be driven out in this way. Once they had been driven out of the country that was familiar to them they would have little chance of surviving the hazards of predators, exposure and starvation. At the beginning of winter there were left in the wood a number of pairs, each with its own territory. The minimum requirement for a territory was a hole in a tree or some other suitable place for making a nest. But when Kluijver artificially increased the number of places for nests by putting nest-boxes in the wood, the number of territories did not increase up to the limit set by the number of boxes. Nor did they depend directly on the amount of food because at the time when they were being established there was a lot more food present than there would be during winter; yet there were rarely enough birds left in the wood to exhaust the supplies of food during winter. The size and number of territories seemed to depend on something else that could not be measured. Like the Australian magpies the great tits seemed to possess innate qualities of 'aggressiveness' or 'unsociability' which largely determined the size of the territories and hence the number of birds left to overwinter in the wood.

During winter the pairs disbanded and the birds disregarded territories. Each bird ranged over a much wider area seeking food and, in a place where food was abundant, a number would congregate without any sign of 'unsociableness'. But each bird would recognize without defending it, a smaller, usually a central part of its 'home-range' in which it would find places for roosting and sleeping, and in which, in due course, it would choose a nesting site. Kluijver called this area its 'domicile'. The domiciles and home-ranges of different birds quite often overlapped and were usually larger than territories which never overlapped.

With the approach of summer, pairs were re-formed and territories were resumed briefly. When the nestlings had to be fed the boundaries of the territories were largely ignored and the birds ranged widely in search of food.

The tits were like the Australian magpies in that the size of the breeding population was determined by the 'territorial' behaviour of the birds. They differed from the magpies in that (*a*) the territories were held by pairs instead of tribes. (*b*) The tits recognized a home-range as well as a territory whereas with the magpie the territory and the home-range coincided. (*c*) The 'surplus' tits that were driven out of the territories were more likely to die quickly than the magpies.

Blair (1951) studied a population of beachmice *Peromyscus polionotus* living in 65 acres of Santa Rosa Island, Florida. He trapped the area thoroughly with traps set out on a grid; he released the trapped mice and watched them as they went back to their holes; he followed tracks in the sand; and he mapped the whole area for mouse-holes and other 'signs'.

Each mouse, as it became mature, established itself in a home-range in which it usually spent the rest of its life. The usual size for a home-range was about 5 acres but they were larger in open places and smaller in places where the vegetation provided a lot of shelter. Each home-range contained about 20 holes of which about 6 would be in general use at one time (Fig. 6.02). When a mouse that had been trapped was released it usually went unfalteringly to a nearby hole if it was a 'resi-

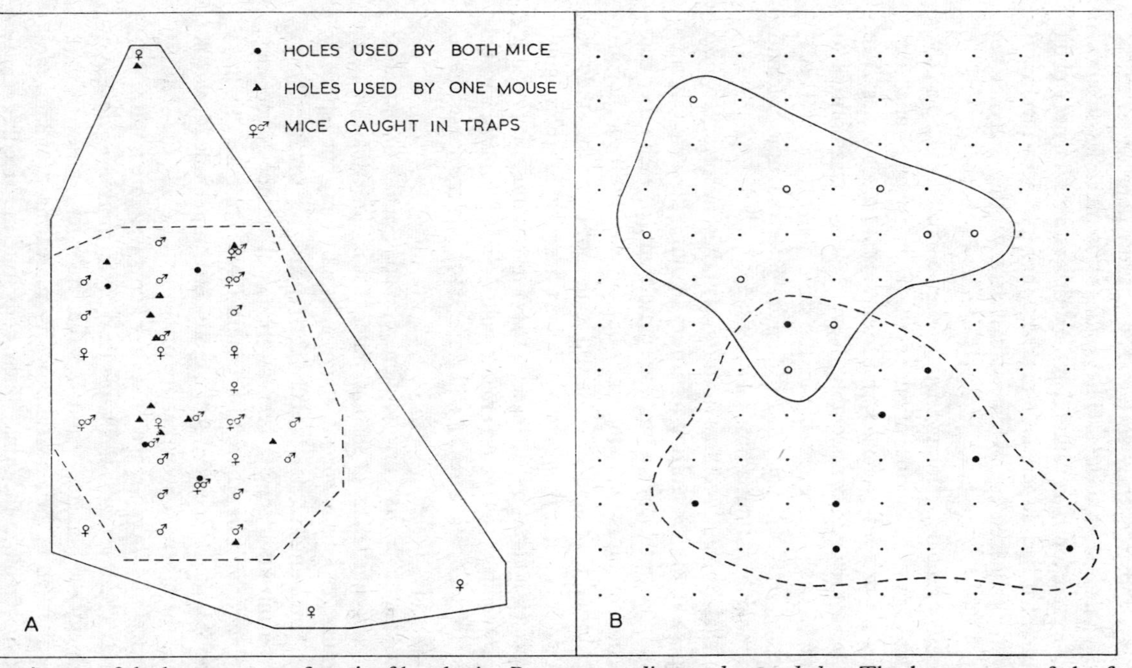

Fig. 6.02. A, map of the home-ranges of a pair of beachmice *Peromyscus polionotus leucocephalus*. The home-range of the female was larger than that of the male. The symbols indicate the places where the mice were trapped and the holes that they used. (Modified after Blair, 1951.) B, theoretical map of the home-ranges of two animals. Note that the home-ranges were mapped by trapping on a grid, and that the home-ranges overlap. This is quite usual for home-ranges as distinct from territories which do not overlap. (Modified after Benton and Werner, 1958)

dent'; but an immigrant that was strange to the area behaved with much less certainty, and frequently sheltered in shallow depressions in the sand or other makeshift shelters. The resident was much less likely to be taken by a predator. Blair found that the death-rate was higher among young adults that had dispersed from where they had been reared but had not yet established home-ranges than at any other stage of the life-cycle.

The density of the population varied between 0·8 to 1·4 mice per acre. Blair estimated that the number of mice on the average home-range area varied from 1·4 to 2·6 for females and from 1·9 to 4·2 for males. The home-ranges overlapped considerably. Nevertheless it was unusual to find two adults of the same sex sharing the same hole. On several occasions Blair observed two males or two females enter the same hole and usually one was driven out again.

It seems that *Peromyscus* differs from *Parus* in that the mice recognize home-ranges without recognizing territories in quite the same way that the birds do. Nevertheless the mice are 'intolerant' of each other and their behaviour towards each other may be considered to be analogous to the 'territorial' behaviour of birds. Blair did not carry the study of 'intolerance' very far in *Peromyscus* but this has been studied thoroughly in other species notably by Errington (1943) in the muskrat *Ondatra zibethicus*, by Chitty (1952, 1957) in the vole *Microtus agrestis*, and by Southwick (1955a, b) and Strecker and Emlen (1953) in the house mouse *Mus musculus*.

The muskrat makes its burrows along the edge of drains, streams, swamps or any other body of shallow water that is close to a supply of grass or grains (especially corn *Zea mays*). At all times, but especially during the breeding season, which corresponds approximately to summer, the muskrats show a keen awareness of numbers, and they will not tolerate crowding beyond a certain limit. So, once this limit has been approached the fiercer or the more determined individuals kill or drive out the others. Those that are driven out, lacking the security of a 'residence' and 'home-range' that they know well, are not likely to live for long. The muskrat is the staple food of the mink *Mustela vison*; but it is chiefly the homeless wanderers that are eaten. Those that are securely established in their own burrow in an area that they know are virtually invulnerable to the mink. These constitute the breeding population. Errington found that they tended to have more and larger litters when the breeding population was sparse than when it was dense. Thus 'territorial' behaviour in *Ondatra* has the same outcome as in *Parus* in that the 'surplus' are driven out into places where they have little chance of surviving; but in addition the 'strife' which is a part of territorial behaviour in *Ondatra* seems also to be reflected in a lowered fecundity.

Chitty showed that the characteristic 'crashes' that recur in natural populations of voles *Microtus agrestis* may not be associated with weather, food, predators or any other obvious component of environment, but only with the numbers of the voles themselves. He found that the voles from a large or declining population reared many fewer offspring than normal; and this reduction in fecundity was also manifest in females of the next generation. Moreover the offspring that were produced at this time seemed more susceptible to the usual hazards that may cause the death of a vole. It had also been noticed that voles taken from a large or declining population often had enlarged spleens. When voles were crowded either in cages or in the field they became increasingly intolerant of each other and fought more fiercely and more frequently. So Clarke did an experiment to test the physiological outcome of fighting, especially of 'losing' the fight. He put a 'strange' vole into a cage where several 'residents' were well established. There was much fighting and the 'stranger' was always beaten and subdued; the residents harried the stranger persistently. After a time Clarke killed the 'stranger' and found that it usually had an enlarged spleen, and adrenals and the thymus was reduced. Dawson (1956) found that an enlarged spleen was often associated with hemolytic anaemia.

The evidence up to this point suggested that the stress that is the outcome of crowding caused physiological disturbances that could be recognized histologically in the spleen, adrenals and thymus and in the population they could be recognized by the abrupt decline in the rate of increase. But when Chitty (1957) tested the voles from laboratory colonies, which had been bred partly from voles that had survived crowding in earlier experiments, he found that many of them had enlarged spleens even though they had been reared uncrowded. This suggested the possibility that the sudden 'crashes' in natural populations may be caused by rather large changes in the frequency of a gene or genes that may be associated with hemalytic anaemia. This idea has still to be tested by experiment.

Strecker and Emlen (1953) and Southwick (1955a, b) confined populations of the house mouse *Mus musculus* in pens 6 feet wide and 25 feet long. In one series of experiments only a limited amount of food was supplied each day. The population increased to the density at which all the food was eaten within two hours of being put out, and then breeding ceased, chiefly because few young ones were born. But the mice maintained their weight. It seemed as if the stimuli provided by the shortage of food and the consequent crowding relative to food caused a change in the physiology of the mice so that food was used for maintaining condition rather than for reproduction.

In another series of pens food was always supplied in abundance, no matter how numerous the mice became. The progress of the population differed rather markedly in the six replicates of this experiment but in five of them the population increased rapidly to a maximum then declined slowly or remained fairly steady despite an excess of food. In all of them there was a high death-rate, approaching 100% at times, in the young mice before they were weaned. Also the number of young that were born decreased markedly after the mice had been crowded for some time.

In one replicate the characteristic results of crowding were observed when there were 15 mice in 150 square feet but in another they did not show signs of crowding until there were 130 mice in 150 square feet. The difference may have been due to the chance occurrence in the first replicate of animals that were excessively 'aggressive'. The outcome of 'territorial' behaviour in these artificial populations of *Mus* was much the same as in the natural populations of *Microtus* and *Ondatra* but the mechanisms were different. There was no sign of enlarged adrenals or spleen.

Territorial behaviour has been recognized in fish, amphibia, reptiles, birds, rodents and in certain ungulates. There are two important features of territorial behaviour. (*a*) The animals recognize a home-range. Probably the chief selective value in this sort of behaviour is that an animal that is well established in a home-range is more secure against predators and other hazards than one that is not established in familiar surroundings. (*b*) The animals display a peculiar intolerance of crowding. Birds may defend actual 'territories' driving the unsuccessful surplus out into unfamiliar country where their chance to survive and multiply may be small. Mammals do not as a rule defend specific territories; with them 'territorial behaviour' takes the form of 'strife' that either leads, as in the birds, to the eviction of the unsuccessful surplus or to physiological changes in the whole population that reduces the rate of increase by one mechanism or another. The net result is that the breeding population lives in home-ranges where their security is great. And their numbers are kept down below the density at which shortages of food, nesting sites or other resources are likely to occur.

A lot has been made of the 'regularity' of the so-called 'cycles' in the populations of rodents, birds and other vertebrates (Dymond, 1947). It has not been established beyond doubt that the fluctuations in any of these populations are truly cyclical (Andrewartha and Birch, 1954, sec. 13.4). When you look at the problem from the perspective that I have indicated in this section it seems to be unimportant whether the fluctuations are cyclical or not. The important point is that the increase of the population is arrested before all the food has been used up, and that, over a long period, the average size of the population remains small relative to its supplies of food and other resources.

The physiological changes that occur in these animals when they are crowded have not yet been explained. Is the explanation to be given entirely in terms of physiology? Or are the physiological changes merely the outward manifestation of genetical changes? If the explanation is genetical must it be given in terms of balanced polymorphism or is there some other mechanism? Whatever may be the answers to these questions the deeper problem of how territorial behaviour evolved remains a puzzle because the individuals that yield have little chance of contributing to the next generation yet the persistence of this behaviour in the species, perhaps the survival of the species itself, depends on the presence of individuals that yield.

6.123 Fluctuations in laboratory populations of invertebrates

In section 6.121 I mentioned that when *Tribolium* was reared with temperature, moisture, food and space kept as constant as possible the population still did not maintain a steady density but fluctuated widely as shown in Figure 6.01. Fluctuations were also observed in a similar experiment with *Calandra* (Birch, 1953a). Many experiments of this sort have been done with different species including *Paramoecium*, *Daphnia* and insects. Although no others, that I know, have been carried on long enough to demonstrate such striking fluctuations as in Figure 6.01, there is evidence that fluctuations may be the rule rather than the exception in this sort of experiment.

It is difficult to explain these fluctuations in terms of the environment which the experimenter has kept as constant as he can. It is also difficult to accept the idea that similar physiological mechanisms are operating to cause fluctuations in insects as in mammals. But it may be more plausible to suggest that an analogous genetical mechanism is operating in both groups. At present this is mere speculation, but it suggests a promising field for research. On the other hand there may be no common principle linking self-regulatory mechanisms in diverse species of animals. With rodents, birds and other 'territorial' species changes in the social structure of the population may be sufficient to explain the fluctuations in density that are observed (Petrusewicz, 1957). With other non-territorial species changing age-structure in the population and 'acclimatization' to crowding may be more important (Ricker, 1954).

6.124 Age-structure in natural populations

Lotka (1925) showed mathematically that the distribution of ages in a population in which the birth-rate and the death-rate for each age-group remain constant and which is increasing in unlimited space would approach the stable age-distribution (sec. 6.1). In nature this is not likely to happen because birth-rates and death-rates do not remain constant nor do populations increase continuously. It happens more often that vicissitudes of weather, food, predators, disease, etc. from time to time destroy all, or most, of the more vulnerable stages in the life-cycle. This happens regularly each winter with certain insects; then, with the return of summer, the population resumes activity with only one age-group present. With vertebrates the vicissitudes may be less pronounced, and the uniformity correspondingly less extreme; but the same principle holds; indeed, students of wildlife set great store on their ability to 'age' the animals that they study.

6.2 OTHER ANIMALS OF DIFFERENT KINDS

In the next few sections I shall describe a number of 'case-histories' in which the animals' chance to survive and multiply has been largely influenced by the activities of other species of animals. The sorts of relationships that may occur are exceedingly diverse but they may be classified into four general categories which are indicated by the headings of sections 6.2 to 6.3.

6.21 *Predators*

6.211 The 'biological control' of insect pests

About 150 years ago the scale insect *Icerya purchasi* was accidentally introduced into southern California from Australia. By 1890 it had spread and multiplied so much that it threatened to destroy the citrus industry. In 1888 a colony of 129 ladybirds *Rodolia cardinalis*, from Australia, were established inside a cage that had been built over an orange tree; the tree was carrying a dense population of *Icerya*. By April

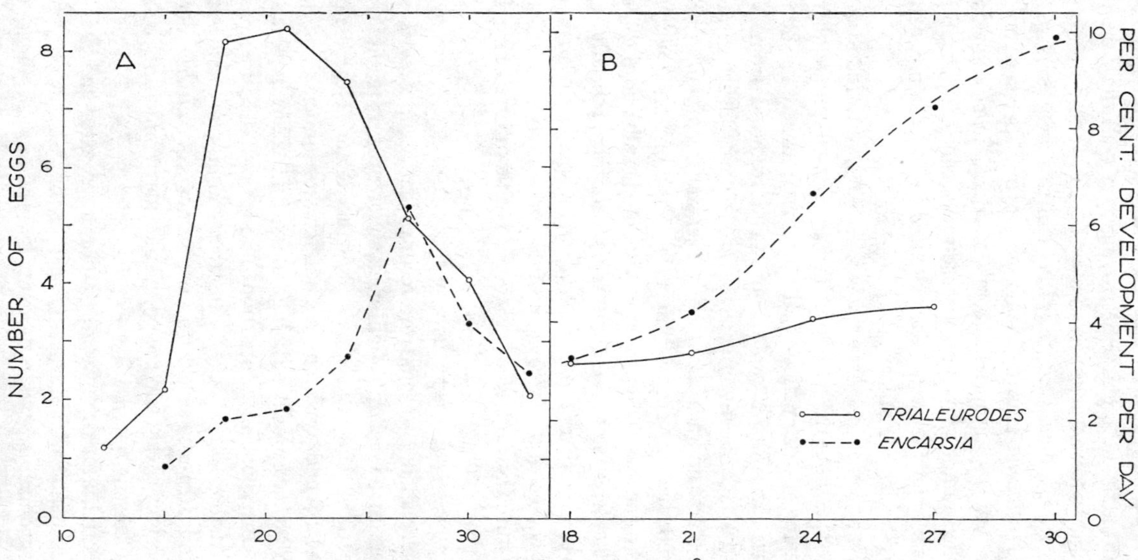

Fig. 6.03. The influence of temperature on A, the daily rate of egg-production, and B, the speed of development of a predator *Encarsia formosa* and its prey *Trialeurodes vaporariorum*. Note that the relative advantage possessed by the prey at low temperature disappears at high temperature. (After Andrewartha and Birch, 1954 – data from Burnett, 1949)

1889 the ladybirds had increased greatly and had destroyed nearly all the scale insects. The cage was removed, and within 3 months *Rodolia* had spread through the orchard destroying nearly all the *Icerya* that were in it. This success was repeated elsewhere in southern California and ever since there have been few *Icerya* in this area. Smith (1939) explained this result in terms of the high powers of dispersal of predator relative to its prey and also the high rate at which *Rodolia* multiplied. They multiplied rapidly not only because their food was plentiful and readily accessible but they were also favoured by other components of environment especially weather.

The woolly apple aphid *Eriosoma lanigerum* was introduced into Western Australia about the middle of last century. Despite regular and frequent spraying with insecticides, the aphids remained an abundant pest until the encrytid *Aphelinus mali* was introduced from North America about 1925. Shortly after *Aphelinus* was established *Eriosoma* became quite scarce and has remained so ever since. From midsummer through winter one may search diligently in the orchards with little chance of finding an aphid. But a few survive and usually, during spring, small colonies can be found, here and there, multiplying briefly before the predators destroy them. Both *Eriosoma* and its predator remain dormant during winter, but *Eriosoma* resumes its activities at a slightly lower temperature than *Aphelinus*. Hence it is able to breed for a brief period during early spring.

With *Metaphycus helvolus*, predator of the scale insect *Saissetia oleae*, the differential response to temperature of predator and prey is more striking. It was introduced into California about 1937. It readily established itself both in the coastal area and farther inland. Near the coast it has caused a large and consistent reduction in the numbers of *Saissetia*. But inland, frosts occasionally kill a high proportion of *Metaphycus* without killing a corresponding proportion of *Saissetia*. In between such catastrophes, *Metaphycus* reduces the numbers of *Saissetia* to a low level; but for a period after each severe frost the numbers of *Saissetia* increase without any serious check, and the population may temporarily become quite dense (Clausen, 1951).

The differential influence of temperature on predator and prey was analysed experimentally by Burnett (1949) for the greenhouse whitefly *Trialeurodes vaporariorum* and its predator *Encarsia formosa*. Figure 6.03 shows the fecundity and speed of development of the two species over a range of constant temperatures. When the two species were reared together in a greenhouse at 18°C *Encarsia* bred so slowly relative to

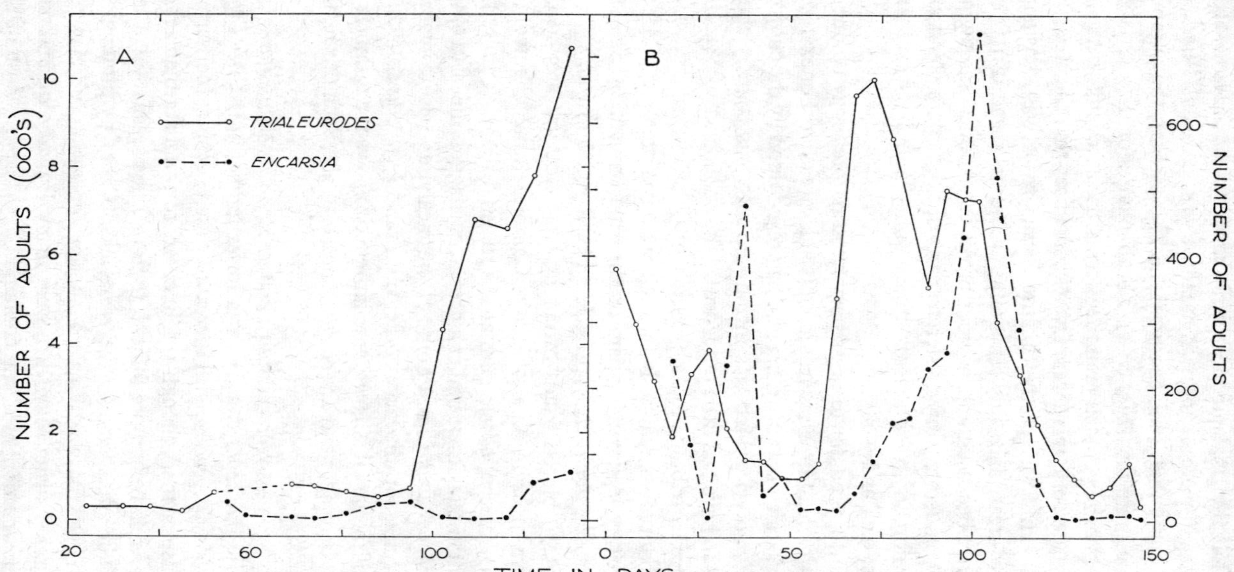

Fig. 6.04. The trends in the numbers of the predator *Encarsia formosa* and its prey *Trialeurodes vaporariorum* when they were reared together in a greenhouse where there were four tomato plants spaced several feet apart. In A the temperature was 18°C; in B, 24°C. (After Burnett, 1949)

Trialeurodes that the predator offered no check to the increase of its prey. When the temperature in the greenhouse was 24° to 27°C *Encarsia* increased so rapidly that it ate nearly all the *Trialeurodes*; the numbers of both became quite few (Fig. 6.04). Local colonies of the prey were exterminated, but the population in the greenhouse as a whole persisted at a low density, because there were always some prey that the predators failed to find. This was partly because *Trialeurodes* was constantly colonizing new areas, and thus remaining one jump ahead of its predator, and partly because the behaviour of the two species was slightly different: they tended to prefer different parts of the tomato plants and to respond differently to gradients in humidity and light (see also sec. 9.12, Figs. 9.03 and 9.08).

The failure of a predator to press heavily on its prey may be due to some other component in its environment besides weather. For example, in Australia the scale insect *Saissetia oleae* is preyed upon by the encyrtid *Metaphycus lounsburyi*, which was introduced into this country about 1925. It seems to have had little influence on the abundance of *Saissetia*, and this has been attributed to the presence of *Quaylea whittieri*, which had been introduced some time before. *Quaylea* is a predator of *Metaphycus*; it must be counted as part of the environment of *Saissetia* because it causes *Metaphycus* to press less heavily on *Saissetia*.

De Bach (1949) placed bands of corrugated cardboard around the trunks of citrus trees in an orchard in California, and counted the mealybugs *Pseudococcus adonidum* that congregated there. Some of the mealybugs contained in their bodies a predatory maggot, the larva of the encyrtid *Anarhopus sydneyensis*. Some of the mealybugs also contained another sort of maggot, the larva of the calliceratid *Lygocerus* sp., which is a predator of the larval *Anarhopus*. Figure 6.05 shows the numbers of *Pseudococcus*, *Anarhopus* and *Lygocerus* counted by De Bach during the summer of 1948.

During early spring *Lygocerus* multiplied more slowly than *Anarhopus*, largely because *Anarhopus* is active at a lower temperature than *Lygocerus*. During May, *Lygocerus* still increased slowly, despite the warmer weather, because its food, though absolutely more abundant than at any other time, was relatively more difficult to find. The explanation of this is as follows. The adult female *Lygocerus* seeks its prey by probing the mealybug with its ovipositor. This is a laborious operation and may take 10 or 15 minutes. During May there were many *Pseudococcus* and only a very small proportion of them contained *Anarhopus*. Most of the mealybugs that *Lygocerus* probed did not provide it with a prey on which to lay an egg. It follows (*a*) that the *Anarhopus* that were present during May stood a good chance of escaping the predator, and (*b*) the predator, having relatively few opportunities to lay an egg, would multiply only slowly. During July the opposite was true. There were fewer *Anarhopus* present in the bands but those that were there were

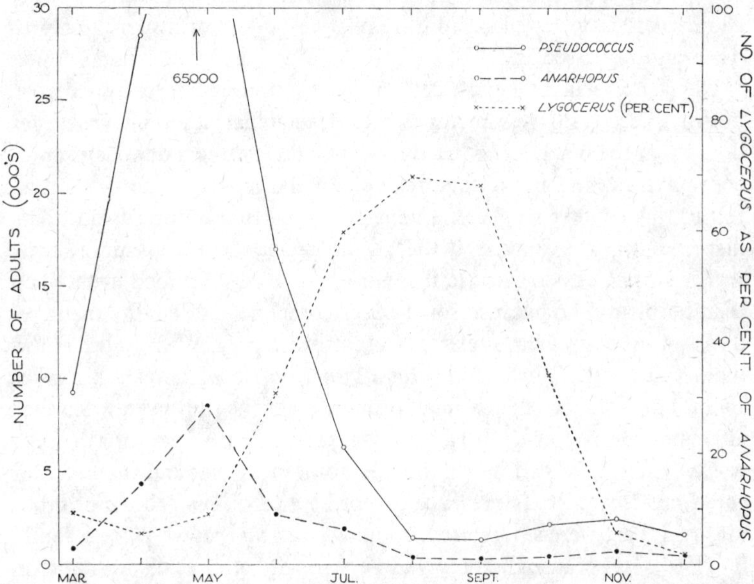

Fig. 6.05. Trends in the natural populations of *Lygocerus* sp., *Anarhopus sydneyensis*, and *Pseudococcus longispinus* in a citrus grove in California. The first preys on the second and the second on the third. (After De Bach, 1949)

very much more accessible to *Lygocerus* because the number of mealy-bugs had decreased greatly. At this time of the year *Lygocerus* multiplied rapidly. In the environment of *Pseudococcus*, *Lygocerus* features as a predator of a predator. The fact that *Lygocerus* presses more heavily on *Anarhopus* when *Anarhopus* is scarce may make it a more important component in the environment of *Pseudococcus* than it would otherwise be (sec. 7.1).

The ladybird beetle *Rhizobius ventralis* was introduced into California about 1895 in the hope that it might 'control' the scale insect *Saissetia oleae*. It became established and has persisted ever since but from the economic point of view it has been substantially a failure. This is because the prey have considerably greater capacity for dispersal than the predators. None of the prey is concealed, in any physical sense, from the predators. Those individuals that escape being eaten and live long enough to grow up and contribute progeny to the next generation do so, not because they are less vulnerable than their fellows, but purely because they are more fortunate: no predator happened to come in time to the place where they were living. In considering such examples, the emphasis is properly placed on the relative powers of dispersal of predator and prey (sec. 4.4).

The failures in biological control have outnumbered the successes, but the successes have sometimes been spectacular, leaving no room for doubt that the newly introduced predator has exerted a dominant influence on the abundance of the prey. The predator that is called 'successful' in this context may be thought of as one that outbreeds and outdisperses its prey. Conversely the 'failure' is one that breeds more slowly than its prey, either because it is not good at finding food even when there is plenty, or because some other component of environment retards its rate of increase relative to that of its prey. With *Encarsia* low temperature (18°C) would do this; with *Anarhopus* it was a predator. Failure to find more than a small proportion of the prey even when they are abundant may be due to a variety of causes. With *Lygocerus* the prey was actually concealed inside the body of a mealybug and the predator could not recognize its presence except by a laborious process of probing; and there were a high proportion of *Saissetia* that were never found by *Rhizobius* because the prey was very much better at dispersing than the predator.

6.212 Predators of vertebrates

The chief predator of the musk-rat *Ondatra zibethicus* in North America is the mink *Mustela vison*. I described the 'territorial' behaviour of *Ondatra* in section 6.122. Each year, as the breeding season approaches, a large proportion of the population is driven out from the familiar country where they have grown up, or at least have been living during the winter. The number of these outcasts is determined by the behaviour of the musk-rat, apparently independently of the presence of predators in the area. The musk-rats that are driven out are almost certain to be eaten by minks and other predators. But those that remain behind, established in a good 'lodge' and living in a familiar home-range, are very much more secure. Errington (1943) considered that this behaviour of the musk-rats largely determined the size of the population; and the predators merely ate those that had already been determined as 'surplus' by their own fellows. And Errington (1946) considered that this principle applied generally to species that manifest 'territorial' behaviour.

Some of the larger ungulates seem to be able to protect themselves from even the canids which Errington classed as the most 'sagacious' and destructive predators except man. They can protect themselves because they are large and because they are organized as a herd. Consequently predators, except man, may have little influence on their chance to survive and multiply. Disease or food may be more important with them.

Some of the smaller ungulates, antelopes, goats, etc., seem to be more susceptible to predators. Some of them live in inaccessible places, as the goats live in craggy hills. Others are more secure when they can form very large herds.

Errington (1946) in his review of predators among vertebrates indicated a number of conclusions. (*a*) Predators are most likely to be influential in determining the abundance of the prey when the prey is a reptile or a fish, or among higher groups when the predator is one of the highly skilful canids (dog, fox). (*b*) Predators are least likely to be influential when the prey manifest 'territorial' behaviour. (*c*) With vertebrates, when predators are important, there is usually also a strong interaction between the two components of environment 'predators' and 'a place in which to live'.

6.213 Obligate and facultative predators

Among the predators mentioned in the preceding two sections *Encarsia formosa* preys on *Trialeurodes vaporariorum* and no other species; similarly, *Aphelinus mali* is restricted to *Eriosoma lanigerum*. For this reason *Encarsia* and *Aphelinus* are called *obligate* predators. There are many obligate predators among insects.

On the other hand, *Rhizobius* will eat other species of scale insects besides *Saissetia oleae*; and most vertebrate predators have a number of species of prey that will serve them as food. Such predators are called *facultative*, to indicate that they are not restricted to one species of prey.

6.22 *Non-predators*

6.221 An alternative prey for a facultative predator

The caterpillars of the moth *Levuana irridescens* used to be a pest of coconut palms in Fiji. During a serious outbreak it was usual to find virtually every palm defoliated over large areas. The outbreaks were extensive because *Levuana* breeds rapidly; and the devastation was complete because the moths show very little tendency to disperse; they usually lay their eggs on a palm that is adjacent to the one on which they grew up.

About 1925 the tachinid *Ptychomyia remota* was introduced into Fiji. The adult *Ptychomyia* lays its eggs on the caterpillars of *Levuana* and the maggots which hatch from the eggs eat the caterpillars. *Ptychomyia* possesses the qualities of a 'successful' predator: it breeds more rapidly than its prey; and its powers of dispersal are incomparably greater than those of *Levuana*. Its success was immediate and thorough: within a year *Levuana* had become, and has remained ever since, a rare insect. Not even the famous example of the introduction of *Rodolia* into California was more spectacular than this. Yet *Ptychomyia* owed its effectiveness to one additional circumstance which was not required for *Rodolia*.

Only caterpillars that have reached a late stage in the third instar are suitable for *Ptychomyia* to lay eggs on. In the absence of the predators, or during the early stages of an outbreak, it is usual to find a substantial overlapping of generations; all stages of the life-cycle are present together including the advanced caterpillars that serve *Ptychomyia*. In the declining stages of an outbreak, when there are not many *Levuana* left, the overlapping of generations tends to disappear, and there comes a time when there are no larvae present in a suitable stage for *Ptychomyia*. The adults of *Ptychomyia* live for no more than 10 days; quite a large proportion of the predators may die without reproducing if the period when no caterpillars of a suitable stage are present should be prolonged for several weeks.

On one island where no alternative food existed for *Ptychomyia* this sequence of events caused the numbers of *Ptychomyia* to fall so low that *Levuana* was able to increase greatly again before the predators 'caught up' with them. Elsewhere in Fiji two other species of caterpillars *Plutella maculipennis* and *Eublemma* sp. that live in the 'bush' adjacent to coconut plantations serve as alternative prey for *Ptychomyia*. The predators maintain their numbers on these alternative prey during emergencies, returning to *Levuana*, which they prefer, when *Levuana* become available again. These species which serve as alternative prey for *Ptychomyia* are important animals in the environment of *Levuana*; in their absence, but with *Ptychomyia* present, *Levuana* might be less abundant in Fiji than it was before 1925, but certainly it would be much more abundant than it is now.

I think that many predators that press heavily on their prey follow the general pattern of *Ptychomyia*; that is, they are facultative predators with a strong preference for a particular prey but they can maintain themselves, during times of scarcity, on alternative species that also abound in the area. A special case is the 'carrier' of a disease that may be fatal to another species. Rats and other small rodents carrying typhus is a familiar example from the ecology of man.

6.222 Non-predators that need to eat the same sort of food, live in the same sort of place or share some other resource

The rabbit, in Australia, is a serious pest of pastures. The presence of many rabbits in an area means that it will carry fewer sheep. The rabbit eats much the same sort of food as the sheep: there are no foods eaten by the sheep that the rabbit will not eat but there may be a few that will serve the rabbit but not the sheep. In arid country the sheep may travel as far as 5 miles away from water for its food but the rabbit rarely moves more than 2 miles away from water in hot dry weather. The requirements of the two species differ in other important ways also, but they overlap sufficiently to make the rabbit an important animal in the environment of the sheep.

In South Australia the scelionid *Scelio chortoicetes*, during its larval stage, feeds on the eggs of the grasshopper *Austroicetes cruciata*. The nymphs and adults of the grasshopper are eaten by a number of species of birds (sec. 6.11). Usually, the grasshoppers are numerous relative to both the birds and to *Scelio*; both sorts of predator consume only a very small proportion of the food that is available to them. But at infrequent intervals drought kills off all the grasshoppers except those that happen to be living in local situations that are moister than elsewhere. Birch and Andrewartha (1941) observed this happen during the drought of 1940. The birds sought out the few grasshoppers that were left in these local situations and ate virtually the lot: scarcely any survived to lay eggs. Consequently food for the larval *Scelio* became exceedingly scarce; they did not become extinct from the area but they were reduced to very low numbers. On such occasions the birds become important animals in the environment of *Scelio*.

These two examples indicate how diverse may be the relationships between animals that need to use the same resource. Andrewartha and Birch (1954) discussed a number of examples. We did not find, in any of the examples that we analysed, two species that had precisely the same requirements. This was not surprising. In order that one species should split into two, it must be divided into two genetically isolated populations for many generations. The chance that their requirements would remain identical during this period is negligible.

It has been shown by Gause (1934), Park (1948), Birch (1953) and others that when two species whose requirements overlap closely are crowded together for many generations in a confined space one species will eventually die out. This result was predicted by the mathematical theory of Lotka (1925) and Volterra (1931) who, quite independently of each other, derived equations for 'interspecific competition'. This theory predicted that no two species having identical requirements could continue to live together in the same place.

This theory continued for a long time to attract much attention but I doubt whether it really has much relevance for the study of natural populations because:

(*a*) No two species are ever likely to have identical requirements (see above).

(*b*) In nature rarity is the rule for most species and intense crowding for any length of time is quite unusual (sec. 9.32).

(*c*) Other sorts of interactions between species not involving crowding or competition are much more important than competition (secs. 6.22, 8.2). Andrewartha and Birch (1954, Chap. 10) discussed these points in greater detail.

6.223 Non-predators whose importance does not depend on sharing the same resource

The two species of blowflies *Lucilia sericata* and *L. cuprina* look very much alike; for many years they formed a pair of 'sibling' species which could not be told apart. But now that they can be distinguished it has become possible to demonstrate striking differences in their physiology and behaviour. Waterhouse (1947) exposed, out of doors, near Canberra, 27 fresh carcasses of sheep at different times throughout the year, and counted the numbers of different species of blowflies that completed their development on the carcasses. A carcass, on the average, produced about 10,000 adult flies of various species, but the variability was great – from 109 to 67,000. These numbers seem small: a carcass weighs about 100 lb.; and in the laboratory, in the most favourable circumstances, it is possible to rear about 7,000 *L. cuprina* from 1 lb. of meat (Nicholson, 1950).

Waterhouse noticed that, at certain seasons of the year, *L. cuprina* laid many eggs on the carcasses while they were fresh. Nevertheless 5 carcasses exposed during March to May yielded no adult *L. cuprina*, and none emerged from 6 carcasses exposed during June to August; the remaining 16 carcasses yielded a total of 108 adult *L. cuprina*.

Waterhouse also collected 26 'fly-blown' sheep, that were still alive, and reared the flies from them. From 16 sheep, *L. cuprina* was the only species that was reared. From the remaining 10, *L. cuprina* was easily the most abundant but there were also a few of a number of other species that are commonly found breeding in carrion. They included *L. sericata*, *Calliphora augur*, *Calliphora stygia* and *Chrysomyia rufifaces*. These species had constituted the majority of those reared from carcasses in the previous experiment.

The following experiment by Mackerras and Mackerras (1944) helps to explain these observations. I describe their experiment in their own words.

> Eight ewes which had soiled or moist crutches were exposed to a pure culture of mature *L. sericata* for a week. Oviposition occurred on two sheep; on one a strike did not follow, on the other on which oviposition had occurred on the 6th day small maggots were present on the 7th day. Several hundred *L. cuprina* were then added to the insectary and within four hours every sheep had egg-masses on the breech. A definite strike occurred on each animal. Some eggs were collected from each breech and were allowed to develop into flies and the curious fact emerged that in seven out of eight both *L. cuprina* and *L. sericata* were present while the eighth contained *L. cuprina* alone. Apparently the act of oviposition by *L. cuprina* had in some way stimulated *L. sericata* to lay eggs on areas which they had previously ignored.

In an area where there are living sheep *L. cuprina* may be an important animal in the environment of *L. sericata* because the presence of *L. cuprina* allows *L. sericata* to breed in sheep as well as in carrion.

In carrion *L. cuprina* is merely an ineffectual and much suppressed sharer of other species' food, but in the living sheep it does much better than all the other species that suppress it in carrion.

Lloyd, Graham and Reynoldson (1940) and Lloyd (1943) studied the ecology of a worm and several species of insects that live in a sewage filter-bed near Leeds. The group of animals comprised:

An enchytraeid worm, *Lumbricillus lineatus*.

Four midges of the family Chironomidae, *Metriocnemus hirticollis*, *M. longitarsus*, *Spaniotoma minima* and *S. perrenis*.

And two moth midges of the family Psychodidae, *Psychoda severina* and *P. alternata*.

The food in the filter-bed consisted of a mass of algae which formed an encrusted layer around the stones near the surface, and a slimy mass of bacteria, Protozoa and fungi in the depths.

The worms *L. lineatus* and the larvae of the midge *M. longitarsus* lived chiefly in the top foot of the bed and fed on the surface layer of algae. Once a year, during the spring, their numbers, especially those of the worm, increased so greatly relative to the growth of the algae that they caused it to disintegrate and settle into the depths of the beds, where it augmented the supply of food for the animals that were living there. In the absence of these 'surface-scourers' the beds would become 'choked' and unsuitable as places for the other species to live. After the surface layer had sloughed off in this way, food became very scarce for *L. lineatus* and *M. longitarsus*. Because the sloughing was chiefly caused by *L. lineatus* these worms were clearly important animals in the environment of *M. longitarsus*; they also needed to eat the same sort of food, but they were chiefly important because they greatly reduced (without themselves consuming very much), the stock of food available for *M. longitarsus*.

The larvae of the fly *M. longitarsus* fed chiefly on the encrusted algae, but in the course of their feeding they destroyed many eggs or small larvae of the other species that happened to be there. This species bred continuously throughout the year, with continuously overlapping generations. So larvae were always present in the surface layers of the beds, but they were usually most abundant from December to March and again in June. The other three species laid their eggs close to the surface, so their chance of survival during this stage depended quite largely on the density of the population of *M. longitarsus*. Accordingly the timing of the life-cycle of the other species relative to the density of the population of *M. longitarsus* was important. Thus *M. hirticollis* had its major flight during June when the population of *M. longitarsus* was usually dense, so the death-rate among eggs of this species was usually high unless some adverse circumstance (perhaps weather) had reduced the numbers of *M. longitarsus*. Similarly for *S. minima*, except that this species, having a shorter life-cycle during summer, had a better chance of recovering from an initial set-back. These two species, and especially *M. hirticollis*, were relatively rare, but occasionally (during the course of these investigations it happened twice in eight years) an unusually hot, dry period during June so reduced the numbers of *M. longitarsus* that the survival-rate among the eggs of *M. hirticollis* and *S. minima* was high, and these species became temporarily quite abundant.

Because *M. longitarsus* did not usually live or feed in the depths of the beds where *M. hirticollis* and *S. minima* lived (once the eggs had hatched), it did not require to share, to any important extent, any of the resources needed by the other two species; on the contrary, the stocks of food for the latter were sometimes augmented by algae which sank into the depths from the surface layers where *M. longitarsus* was living. Nor is it strictly a predator although it may eat some of the eggs that it destroys. It is an important animal in the environments of *M. hirticollis* and *S. minima* chiefly because it destroys their eggs.

The larvae of the two moth midges *P. severina* and *P. alternata* lived in the depths of the beds, and the adults probably always laid at least some of their eggs in the depths, thus avoiding the hazards associated with the presence of *M. longitarsus* near the surface. They had to share their food with *M. hirticollis* and *S. minima*. Moreover, these species interfered with them and destroyed them in much the same way as they themselves were destroyed by *M. longitarsus*. So *M. hirticollis* and *S. minima* were important animals in the environments of *P. severina* and *P. alternata* partly because they needed to eat the same sort of food and partly because they were directly destructive.

6.3 PATHOGENS

The larva of the small tineid moth *Plutella maculipennis* lives and feeds on the leaves of cabbages and other crucifers. Ullyett (1947) counted, at weekly intervals during 1938/9, the *Plutella* that were living on cabbages in a garden near Pretoria. He also counted the number that had been eaten by predators, or had died from the disease that is caused by the fungus *Entomophthora sphaerosperma*. The results are shown in Figure 6.06.

Widespread outbreaks of the disease occurred during February 1939 and again in November 1939. During the first outbreak the population of *Plutella* was reduced to about 2 per plant and remained at this low level for about 10 weeks. Precise counts were not made during the second outbreak but Ullyett wrote that the disease had followed much the same course as before. Between outbreaks very few larvae died from this disease. But the fungus must have been present and widespread, because outbreaks of disease occurred whenever there had been enough rain to keep the leaves continuously wet for 3 or 4 days. Light continuous rain was more effective than heavy intermittent rain; even continuous mist would do. The disease was observed to break out with equal severity in a number of different populations of different densities, and Ullyett concluded that the onset of the disease was independent of the density of the population at the beginning but depended only on the weather. Once the rain abated and the leaves of the plants remained dry for several days the disease came to an end.

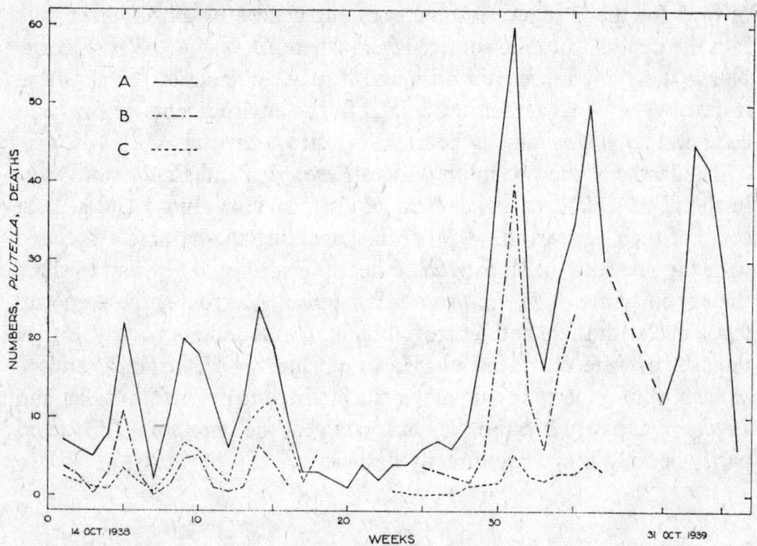

Fig. 6.06. A, the numbers of *Plutella maculipennis* in a garden near Pretoria; B, the deaths due to facultative predators; C, deaths due to obligate predators. (After Ullyett, 1947)

Ullyett considered that between outbreaks of disease (e.g. during the first 15 weeks shown in Figure 6.06) predators prevented *Plutella* from increasing to very large numbers. During an outbreak of disease the predators became very few because there was so little food for them. Consequently *Plutella* multiplied rapidly for a period after the disease had abated. But even at these times *Plutella* did not become numerous enough to eat a large proportion of its total stocks of food.

It was the interaction between weather, disease and predators that determined the abundance of *Plutella* during the period of this investigation – and the most basic variable was the frequency and severity of spells of wet weather.

In Canada extensive laboratories have been built expressly for the study of diseases of the insect pests of forests. One method, that has given promising results, is to 'mass-produce' the pathogens in the laboratory and then spray them on to the trees where the pests are numerous (Cameron, 1950).

6.4 THE ECOLOGICAL WEB

In section 6.223 I described the ecology of some of the animals that live in a sewage filter-bed near Leeds. The chironomids *M. hirticollis* and *S. minima* eat the same food as the psychodids *P. alternata* and *P. severina* and all live together in the depths of the beds. The abundance of the chironomids depends on the abundance of *M. longitarsus* in the surface layers, which is influenced by the numbers of *L. lineatus*. Both *M. longitarsus* and *L. lineatus* form part of the environments of the psychodids even though they live in a different place and do not influence them directly.

In section 6.211 I described how *M. lounsburyi* pressed less heavily on its prey *S. oleae* because it was itself preyed on by *Quaylea*. Similarly *Lygocerus* is part of the environment of *Pseudococcus* because it is a predator of a predator of *Pseudococcus*.

The sheep-blowfly *Lucilia cuprina* costs the farmers in Australia many millions of pounds a year, and much effort has been put into finding ways to make it less abundant. In section 6.223 I mentioned that in nature, very few *L. cuprina* breed in carrion because they have little chance of surviving when they are crowded by *L. sericata*, *Calliophora augur*, *C. stygia* or *Chrysomyia rufifaces*. These are called 'secondary flies' and *L. cuprina* is called the 'primary fly' because it breeds chiefly on the living sheep.

There are three small wasps that prey on the blowflies; *Mormoniella vitripennis* and *Stenosterys fulvoventralis* have been in Australia for a long time, but *Alysia manducator* was especially introduced from Europe about 1928. In so far as these predators reduce the amount of crowding in carcasses the chief result of their presence will be to increase the abundance of *L. cuprina*. Fortunately none of them can outbreed or outdisperse its prey, and they make very little difference to the numbers of any sort of blowfly.

About the same time as these predators were being bred artificially and spread about the countryside a lot of effort was also going into the perfection of traps for catching the adult flies. And farmers were urged to use this method to 'control' the sheep blowfly. The traps caught more secondary flies than *L. cuprina*. Their chief effect was to reduce the crowding in carcasses; so they merely served to increase the abundance of the sheep-blowfly. If a scientific study of the ecology of *L. cuprina* had been made first, these mistakes might have been avoided and a lot of money might have been saved.

The wasps *M. vitripennis*, *S. fulvoventralis* and *A. manducator* are predators of *L. cuprina*, as well as of the 'secondary' flies, but they go mostly to carrion; so they are chiefly important in the environment of *L. cuprina* as predators of the other species that prevent it from breeding in carrion. Similarly the men who trapped the 'secondary' flies became part of the environment of *L. cuprina* because they destroyed, without preying on them, some of the 'secondary' flies, and thus increased the opportunity for *Lucilia cuprina* to multiply.

At first it may seem that there is a bewildering diversity in the sort of indirect relationships that may occur between an animal and the other animals of different sorts in its environment. But they can all be fitted into the same pattern of predators and non-predators which I described in the preceding sections. Figure 6.07 summarizes all the relationships that may occur. Of course, in nature, no one species is likely to have all these categories represented in its environment; some species will have none that is important, and few will have more than one or two. But I have drawn them all in one diagram in order to illustrate the principle.

The diagram shows a grub (a larva of a beetle) in the centre of the lower circle, at position A. It is in the centre with all the arrows leading towards it because it is the animal whose environment is being analysed. This imaginary grub eats the leaves of thistles which grow as weeds in a pasture in which the chief plants are grasses and clovers. The arrow *a* comes from a caterpillar that also eats the leaves of thistles; it is a non-predator that needs to share the same resource. The arrow *b* leads from a sheep that does not eat thistle but in seeking its own food of grass and clover it tramples on the thistle and on some of the grubs; it is a non-predator whose importance does not depend on sharing a common resource. The arrow *c* leads from a bird that eats the grub; it is a facultative predator because it also eats the grasshopper (which is drawn at position *d*) during winter when the grubs are not present. The grasshoppers are an alternative prey for a facultative predator.

Each one of these animals may have in its own environment animals that may come into any of these same four categories. This is indicated by the arrows coming in from the upper circles in Figure 6.07. Third and higher levels are logically possible but are not of much practical importance.

If we place *Lucilia cuprina* in the centre of the first circle, at position A, then *L. sericata* is at position a and its predator *Alysia* is immediately above at c' and the man with his trap is in the same circle at b'. If the psychodid *P. alternata* be placed in the centre of the lower circle then *M. hirticollis* is at a, *M. longitarsus* is immediately above at b' and *L. lineatus* is at b'' – immediately above *M. longitarsus* at b' (this third level of complexity has not been drawn in the diagram). It so happens that the abundance of *M. longitarsus* is influenced more by weather than by

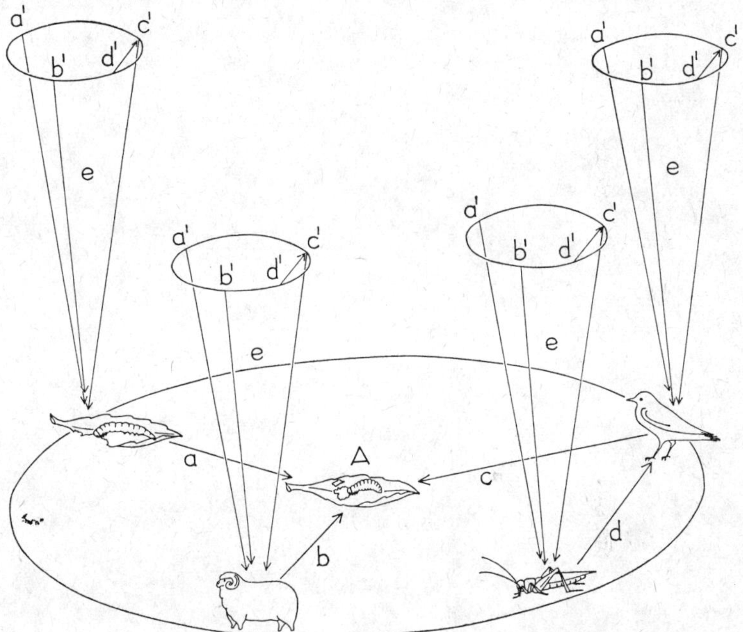

Fig. 6.07. A diagrammatic representation of the 'ecological web'. For explanation see text. (After Andrewartha and Birch, 1954)

L. lineatus but I am ignoring all other components of environment in this diagram because I want to represent merely the *ecological web* which comprises all the possible relationships between an animal and the other animals of different sorts that may form part of its environment.

Each arrow in Figure 6.07 defines a 'niche' (sec. 1.0). But there is a place in the diagram only for those arrows that lead to the central animal at A. When the ecology of this species is understood other species from the community may be studied. If a deeper knowledge of the community is needed this is a more practicable way to build it up than the direct approach in which an attempt is made to study the whole community at once.

6.5 FURTHER READING

Andrewartha and Birch (1954), Chapters 9 and 10. Lack (1955) expounded a different point of view from that which I discuss in section 6.122.

CHAPTER VII

Components of Environment: Food

7.0 INTRODUCTION

I mentioned in section 6.211 that the woolly aphis *Eriosoma* became quite scarce in apple orchards in Western Australia after its predator *Aphelinus* had been introduced to that area. The few *Eriosoma* that now live in the orchards are surrounded by easily accessible food in great abundance. This is the usual condition of most herbivorous animals and of many carnivorous ones also: some component of environment other than food presses on them so heavily that they are quite unlikely to become numerous enough to press on their stocks of food (sec. 9.32).

The converse also happens. Certain species seem to have the capacity to multiply right up to the limit of their food-supply. Food then becomes an important component in their environments (sec. 7.2).

A third possibility is that food may be plentiful yet inaccessible – or (what amounts to the same thing) the animals may not be much good at finding it. In these circumstances their chance to survive and multiply may depend quite largely on whether they get enough food despite the fact that a large proportion of the food in the area usually remains uneaten. Andrewartha and Birch (1954) referred to this paradox of scarcity amidst plenty as 'the relative shortage of food' (sec. 7.1).

7.1 THE RELATIVE SHORTAGE OF FOOD

The ticks *Ixodes ricinus* that live in the rough upland pastures of northern England feed mostly on the blood of sheep and only to an unimportant extent on that of wild animals (Milne, 1949, 1951). During most of the year the ticks remain inactive in the mat of vegetation near the ground. During spring they emerge from this shelter and climb up an exposed grass-stem. If, during the 9 or 10 days that they can survive in this exposed position, they are picked up by a sheep as it brushes by, they cling to the sheep until they are engorged with blood. Then they drop from the sheep and crawl into the mat of vegetation again. They digest their meal and develop to the next stage of the life-cycle. Each tick requires to feed three times during its life – as a larva, as a nymph and as an adult. After the third meal the eggs are laid.

On one farm on which Milne worked he estimated that a tick had about 40% chance of being picked up by a sheep in any one year. The probability that a tick would be picked up 3 times during its life can be estimated as $0{\cdot}4^3 = 0{\cdot}06$. In other words a tick might have a 6% chance of getting enough food to allow it to complete its life-cycle and lay eggs. This farm carried about one sheep to the acre. Had there been more sheep the ticks would have had a greater chance of getting enough food because a tick's chance of being picked up depended only on the number of sheep in the area and their activity. It did not depend at all on the number of ticks in the area because the feeding of the few that did get a meal made no difference to the number of sheep in the area. In other words there was a severe shortage of food in the relative sense but no shortage whatever in the absolute sense.

A striking demonstration of the operation of this principle was provided by the 'Shinyanga Game-extermination Experiment' (Potts and Jackson, 1953). This exercise was aimed at the extermination of three species of tsetse flies *Glossina morsitans*, *G. swynnertoni* and *G. pallidipes* from about 600 square miles of bush.

The tsetse fly sucks blood chiefly from the larger species of ungulates. It requires to feed frequently, especially during hot weather, but it does not stay on or near its host after taking a meal; so each meal is preceded by an independent search for a host. If hosts are few or sparsely distributed a fly may have little chance of getting enough food to produce many or any offspring before it dies, despite the fact that the amount of blood in even one antelope would be more than enough to support many flies. Jackson (1937) argued from these facts that there was likely to be a particular density of hosts that would enable the flies to find food just often enough to maintain births equal to deaths and thus keep the population steady. But if the population of hosts remained sparser than this the deaths among tsetse flies would be likely to exceed births and, after a while, the population might die out.

Accordingly at Shinyanga 50 to 100 hunters were employed to shoot as many as possible of the larger mammals which might serve as hosts for tsetse flies. At the beginning of the experiment there were about 10 such animals to the square mile. During the next 5 years about 8,000 animals were shot, including many that came in from outside the area. Towards the end of this period the numbers of the more important species had been greatly reduced but there was still a small herd of elephants and there were many of the smaller species of ungulates still living in the area. Altogether the animals still living there at the end contained many times more than enough blood to support all the tsetse flies that had been living there at the beginning. Nevertheless the animals were few enough and distributed sparsely enough to give the tsetse flies only a small chance of finding enough food. By the end of the exercise the authors were satisfied that *G. morsitans* and *G. swynnertoni* had been exterminated and *G. pallidipes* had either been exterminated or reduced to extremely low numbers.

The amount of food in the area was independent of the number of flies because those few that did find food did not, by virtue of their feeding, make any difference to the amount or distribution of the food. Also, the flies' chance to get enough food to survive and multiply was independent of the size of the population of flies. With *Glossina* as with *Ixodes* there was no shortage of food in the absolute sense but the relative shortage of food was so severe that the population became extinct.

In section 1.1, I mentioned that in south-eastern Australia the distribution and abundance of *Thrips imaginis* can be very largely explained in terms of weather. To a certain extent weather influences the thrips' chance to survive and multiply directly because dry weather during summer increases the likelihood that the pupal stage will die from desiccation, and cool weather during winter reduces fecundity and retards development. But the weather is chiefly important for its indirect influence on the supply of food for the thrips. In order to explain how this happens it is necessary to describe the life-cycle of *T. imaginis*.

The nymphs and adults live almost exclusively in flowers; they suck the sap from stamens, pistil and petals but in particular they suck the contents of the pollen grains. Pollen is essential; without it they do not grow or lay eggs. The numbers fluctuate regularly with the seasons: they reach a maximum about the end of November (southern spring) and are usually at a minimum during February–March (late summer) and during July–August (winter). The fluctuations are large: in November I have counted as many as 2,000 in a sample of 20 small roses but in July or August it is usual to find about 50 in a similar sample.

The eggs are laid in the tissues of the flower. At 20°C the eggs hatch in about 4 days. The nymphs feed on the tissues of the flower for about 6 days; then they crawl into some sheltered place usually among the debris on the ground and pass through two non-feeding 'pupal' stages which take about 5 days. At lower temperatures all stages take longer. The newly emerged adults fly in search of a place to live and lay their eggs. Their chance of finding one depends on the number and distribution of flowers in the area.

The number of flowers in the area fluctuates regularly with the seasons. In southern Australia the weather resembles that of the Mediterranean region of Europe. During the long, dry summer, annuals wither and perennials produce few flowers. The winter is the time for growth especially of annuals but still there are few flowers. But in the spring there is a great profusion of flowers not only on the numerous and widespread annuals but on many perennials as well. A thrips, taking wing at this time of the year, could scarcely fail to find a suitable flower in which to live for a while and lay its eggs. The survival-rate and the birth-rate are high and the numbers increase rapidly. But the flowers increase even more rapidly so that, even with the expanding number of thrips, there is little chance of an absolute shortage of food except perhaps locally for a brief period.

But the period of great abundance is short-lived. The change that comes with the summer is dramatic. Food becomes scarce and remains so, no matter how little is eaten, because the weather has changed. Flowers that might provide suitable food and breeding places for *Thrips imaginis* are few and widely scattered. A thrips that takes to the wing in these circumstances has little chance of finding food. Many of them die without reproducing. There is still no absolute shortage of food because our samples at this time of the year showed only two or three thrips per rose; and any other flower that was examined showed thrips present at similar low densities. The population which was large in November declines rapidly as summer advances and continues, with perhaps a minor respite in the autumn, to decline through winter until the trend is reversed with the onset of spring (Fig. 7.01).

The great seasonal changes in the abundance of the thrips are related to the seasonal changes in the accessibility of their food. At no time is there an absolute shortage of food but at certain seasons of the year food becomes inaccessible (i.e. unlikely to be found) merely by becoming sparsely distributed (Davidson and Andrewartha, 1948a, b).

Criddle (1930) found that in Manitoba the numbers of grouse increased during outbreaks of grasshoppers, chiefly because the grasshoppers provided abundant food for the grouse when they were nesting. Outbreaks of grasshoppers tend to develop during warm, dry weather, and often come to an end after a spell of cool, damp weather. But the numbers of grouse declined during prolonged cool weather even before the numbers of grasshoppers had changed very much. This was because the grasshoppers were less active during cool weather and the grouse had greater difficulty in finding or catching grasshoppers when they were inactive. In these circumstances many nestlings died from lack of food. There was a relative shortage of food that had been caused, in the first place, by the influence of weather on the behaviour of the grasshoppers.

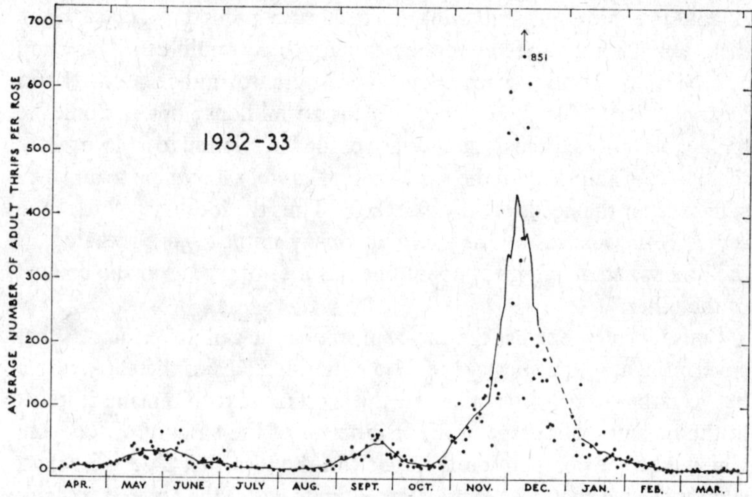

Fig. 7.01. The annual cycle in a natural population of *Thrips imaginis*. The points represent daily records. The curve is a 15-point moving average. (After Davidson and Andrewartha, 1948a)

A relative shortage of food may have been an important component in the environments of primitive men. The following quotation from Braidwood and Reed (1957) makes this point.

> The main point here is that, as Deevey (1956) has already shown, the population of the primary hunting and collecting culture is not limited by the potential amount of food, but by man's ability to get it during the leanest season. Proto-Maglemosian man was probably often hungry, even starving, in country which should have indefinitely maintained a population at least 100 times (10,000%) as high (or 38 people to the square mile). That no hunting and gathering culture ever approached this population optimum . . . is an indication of the human technological level of the hunting-collecting level.

It is the usual condition of most species in nature that they are scarce relative to the total stocks of their food (sec. 9.32). Often this may be explained by the influence of some other component of environment such as weather, predators, disease, etc.; but quite often the chief explanation lies, as in the examples described in this section, in the paradox of scarcity amidst plenty – the relative shortage of food.

7.2 THE ABSOLUTE SHORTAGE OF FOOD

It would be meaningless to think about shortage of food in a place where there was neither any food nor any animal; there would be no question of food shortage in a place where there were 10 million animals and plenty of accessible food to provide for 20 millions; but it would be highly relevant to discuss the shortage of food in relation to a place where there were 5 animals and the food, though entirely accessible to any, was sufficient for the needs of only 2 of them. Thus the idea of too little food is merely the converse of the idea of too many animals – one idea implies the other; and a statement about one has meaning only in the context of the other.

Consequently, one has the choice of stating the general principles that govern the absolute shortage of food either in terms of the competition between the animals (in which case the shortage of food remains implicit in the meaning of competition) or in terms of the amount of food – in which case the competition is implicit in the idea that food is in short supply. I have chosen to express myself in terms of the amount of food; I give my reasons for not making more use of the term 'competition' in sections 9.32, 9.34.

Crowding with respect to food (and hence an absolute shortage of food) may be observed in nature (*a*) when pests such as insects multiply in our fields or gardens without artificial check to their multiplication. (*b*) When insects are used 'successfully' for the 'biological control' of weeds as *Cactoblastis cactorum* was used to control prickly pear in Queensland (sec. 7.221). (*c*) When a predator presses heavily on its prey as happens when a 'parasite' has been used to control an insect pest. In recognizing the economic importance of such species we should not forget that they represent a small minority of the species in nature. It is the lot of the great majority of species to remain rare relative to their food and other resources. (But see also sec. 7.221, last paragraph.)

7.21 *The principle of 'effective' food*

Ullyett (1950) placed varying numbers of eggs of the blowfly *Lucilia sericata* on limited quantities of meat. When there was plenty of food all the maggots grew quickly and reached a weight of about 60 mg. before pupating. When food was scarce they grew more slowly and did not grow so big; but they still pupated provided that they attained the weight of 25 mg. Nicholson (1950) allowed varying numbers of *L. cuprina* to lay eggs on 10 gm. of meat and counted the adult flies that emerged in the next generation. When few eggs were laid on the meat few large flies emerged; when many eggs were laid on the same amount of meat larger numbers of smaller flies emerged; a maximum was reached when 25 females were allowed to lay eggs on 10 gm. of meat. As the number of eggs increased beyond this point the number of adults that emerged decreased until, with 150 females laying eggs on 10 gm. of meat, none emerged, because no maggot got enough food to grow to the minimum size for pupation. All the food was eaten but all was wasted because none was used to contribute to the next generation. But at slightly lower levels of crowding a few maggots were able to grow big enough to pupate. The food that they had eaten may be called 'effective' in the sense that it had contributed to the next generation.

In the very simple situation in which the food happens to be a predetermined amount of non-growing material, say 10 gm. of carrion, the relationship between the number of feeding animals and the proportion of 'effective' food may be expressed by a curve like the one that I have drawn in Figure 7.02 and labelled 'dead food'. The proportion of the total of food that is eaten, including that which may be 'wasted', is represented by the broken line which I have labelled 'A'.

In the more complex situation in which the food is a living plant or animal or a population of living plants or animals it is not realistic to think of the original stock of food being supplied as a predetermined amount which is independent of the number of feeding animals. It is necessary to take into account the increment due to growth that occurs during the time that the feeding animals take to complete a generation. This increment is likely to be small when there are many animals feeding and larger when there are few. If we consider, for the sake of simplicity, the total stock of food that might be available to the feeding animals during the course of one generation, the difference between dead and living food may be represented by curves like those drawn in Figure 7.03. In considering the relationship between the number of feeding animals and 'effective' food when the food is living it is necessary to take into account the slope of the line labelled 'living food' in Figure 7.03. I have tried to do this in drawing the curve labelled 'living food' in Figure 7.02.

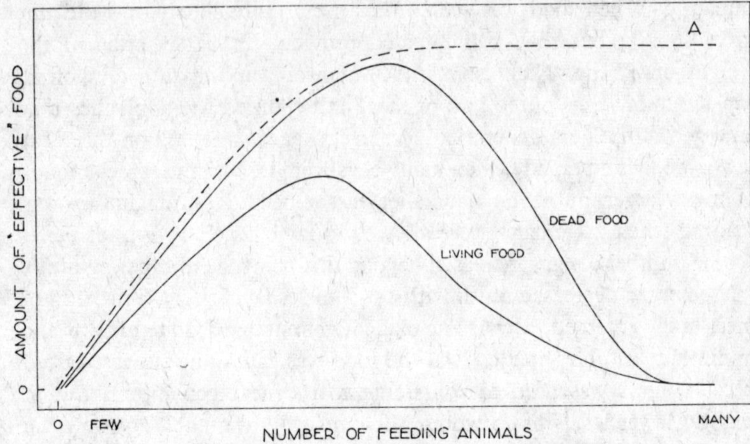

Fig. 7.02. Theoretical curve to show the relationship between the intensity of crowding and the proportion of the total stock of food which is likely to be 'effective' – in the sense that it may be used in the production of progeny for the next generation. For further explanation see text. (After Andrewartha and Birch, 1954)

7.211 Exceptions to the principle of effective food

The codlin moth lays its eggs on the outside of an apple. The larva bores into the apple and, after a brief sojourn near the surface, tunnels to the core where it eats the seeds. In a dense population many eggs may be laid on the one apple and a number of larvae may penetrate the same fruit. But, nearly always, only one survives to reach maturity; usually it is the one that reached the core first. Because it is usual for one but not more than one larva to survive in an apple, there is little chance that any food will be 'wasted' in the way that food was 'wasted' when the population of blowfly maggots increased beyond the 'optimum' density. So the idea of 'effective' food does not apply to the larvae of codlin moth or to other species with similar adaptations for life in a crowd.

Perhaps the most important exceptions are in the group that entomologists call 'internal parasites'. This means a species that lives, during one stage of its life-cycle, inside the body of its host and usually destroys its host. The adult of an internal parasite usually lays its eggs on or inside the body of the host. Some species, e.g. *Telenomus ullyetti*, seem to have an infallible instinct which enables them to avoid placing a second egg inside a host that already contains an egg or a larva (Jones, 1937). But many species lack this instinct or have it only weakly developed. Even so, it is unusual for larvae of internal parasites to compete for food on

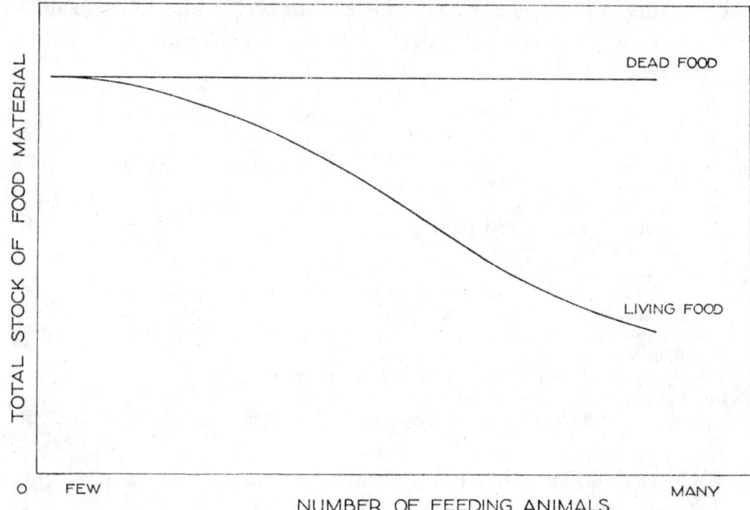

Fig. 7.03. Theoretical diagram to show how the density of the population may influence the total stock of food. The stock of dead food is independent of the number of feeding animals. But the quantity of living food is likely to be less, the more heavily it is grazed or preyed upon, because in these circumstances it is likely to grow less. (After Andrewartha and Birch, 1954)

quite such equal terms as do the larvae of *Lucilia*. If they are crowded they may all feed for a while, but in the end, the weaker are eaten by the stronger, so that the original stock of food is used efficiently to produce its usual quota of individuals for the next generation.

To say that there is an absolute shortage of food is merely another way of saying that the animals are crowded relative to their food (sec. 7.2). The various conditions of crowding may be analysed by classifying the way that the food (or any other resource) is utilized by the animals (Table 7.01). Food may be said to be 'expendable' when the feeding by the present generation is likely to reduce the amount of food for future generations (e.g. the deer on Kaibab Plateau, sec. 7.223: Fig. 7.03, living food). By contrast food may be said to be 'non-expendable' when the amount of food for future generations is independent of the amount of feeding done by the current generation (e.g. blowfly larvae feeding on carrion, or codlin moth, sec. 7.211). Whether food is expendable or non-expendable, the animals' utilization of it may be classified according to whether the principle of 'effective food' is relevant or not. Those who prefer to analyse crowding in terms of 'competition' will find this idea (competition) most relevant to the first cell of Table 7.01, less so to the second and third cells and not relevant at all to the last cell (sec. 9.32).

TABLE 7.01

a. Food is expendable.
  b. Principle of 'effective food' holds:
    e.g. deer, *Cactoblastis* and indeed most herbivorous animals; also fox, ladybirds and indeed most predators except insects that are 'internal parasites'.
  bb. Principle of 'effective food' does not hold:
    e.g. most insects that are 'internal parasites', and a few special cases amongst herbivorous ones.

aa. Food is non-expendable.
  b. Principle of 'effective food' holds:
    e.g. animals feeding in carrion, logs, litter, etc.
  bb. Principle of 'effective food' does not hold:
    e.g. food for codlin moth (nesting sites for many insects, birds, etc.).

7.22 *The circumstances in which an absolute shortage of food is likely to occur in nature*

Whenever the animals in a natural population become short of food (in the absolute sense) this implies, almost by definition, that no other component of environment, including shortage of food in the relative sense, is important. When food becomes short in the absolute sense it often happens that much of the food is destroyed; whole populations of the plants or animals that constitute the food may be wiped out, though not, as a rule, before some individuals have moved off, with the chance of founding new colonies elsewhere (secs. 4.4, 4.5). In nature the complete destruction of a source of food rarely happens, except locally. When a species is found to be responsible for the destruction of a great proportion of its food over broad areas, investigation usually reveals that the eating species has high powers of dispersal and rate of increase relative to those of the eaten species: and the usual picture is a high rate of destruction of food going on concurrently in most of the local situations where the food is distributed. The species that can press heavily on their food like this include many that are of economic importance to us: when their food is one of our valued crop plants we call them pests; when they feed on a weed or a pest we speak of them as effective agencies of 'biological control'.

7.221 The biological control of weeds and pests

Prickly pear is the name given to several species of *Opuntia* that were serious weeds in north-eastern Australia. They were originally brought from America and planted in gardens. They escaped from the gardens and spread into agricultural and pastoral land. Dodd (1936) reported that in 1925 some 30 million acres of potentially good agricultural land was so densely covered by *Opuntia* spp. that it was valueless, and there was in addition about 30 million acres infested rather less densely. The area occupied by *Opuntia* was increasing by about a million acres a year. About this time the moth *Cactoblastis cactorum* was introduced from America and during the next 5 years 3,000 million of them were bred and set free in the area. They multiplied and spread, completely destroying the prickly pear wherever they found it. Dodd estimated that by 1934 the amount of *Opuntia* growing in Queensland had been reduced to one tenth of what it had been in 1925. The subsequent history of *Cactoblastis* in Queensland was well described in the passage from Nicholson (1947) that I quoted in section 4.5.

The end result is that *Opuntia* is now relatively scarce over vast areas where it used to be inordinately abundant; and *Cactoblastis* is correspondingly scarce compared to what it used to be during the period 1925–34. Its numbers are determined chiefly by the amount of food in the area. A complete explanation of the amount of food in the area would, of course, have to take into account all the other important components in the environment of prickly pear as well as *Cactoblastis* (Andrewartha, 1957). This is a point that has been overlooked by some ecologists (sec. 9.34).

In sections 6.211 and 6.221 I described how the predators *Rodolia* and *Ptychomyia* press heavily on their food *Icerya* and *Levuana*. (See also the quotation from Cockerell in section 4.4.) The relationship between these predators and their food differs in details, one from the other and each from *Cactoblastis*, but the general pattern of the relationship is the same in each instance. The food is distributed patchily in small temporary colonies. The colonies usually have a short life because the highly dispersive predators are likely to find them soon after they have been started, and having found them, are likely quickly to destroy them. In each local situation where the predators find a colony of the prey the predators at first increase in numbers and later decrease to zero under the influence of an absolute shortage of food.

If one takes a broader view and considers the large area in which the local colonies are scattered it is clear that the dispersiveness and searching ability of the predators have to be taken into account – which introduces the idea of relative shortage of food. *Cactoblastis*, *Rodolia* and *Ptychomyia* are so good at dispersing and searching that they greatly reduce the abundance of the species that they feed upon. Other species that are not so good at dispersing and searching, yet are good at destroying the local colonies once they have been found (e.g. *Rhizobius*, sec. 6.211), may experience an equally severe relative shortage of food without depressing so severely the abundance of their food (Andrewartha and Browning, 1961).

7.222 Outbreaks of pests

Many of the common insect pests of our farms and gardens have the same potentialities as *Cactoblastis*. Indeed the story of the vine aphid *Phylloxera vastatrix* is like that of *Cactoblastis* except that it has a different ending. This aphid was accidentally introduced into France from America through the port of Bordeaux about 1865. Like *Cactoblastis* it spread quite rapidly and its spread was assisted by man (though inadvertently in this instance); and it killed out whole populations of its host. It destroyed 3 million acres of vines in France before someone discovered that it could be checked by grafting European vines on American root-stocks.

The woolly aphis *Eriosoma lanigerum* has a similar potential with respect to its host the apple; but it is nevertheless quite scarce wherever apples are grown. Apples are now universally grafted on resistant root-stocks; so *Eriosoma* does not thrive on the roots; above ground it is preyed on heavily by *Aphelinus* (sec. 6.211). From time to time outbreaks of other pests such as locusts, thrips, scale insects and army worms remind us that they too, in the absence of other checks, would be able to multiply until they were checked by an absolute shortage of food.

I mentioned in section 7.211 that the codlin moth is an exception to the principle of effective food. It is unusual also in that, no matter how numerous it may become, it does not, by virtue of its feeding, reduce the total stock of food either for the present, or for subsequent generations. This is true because the feeding of the caterpillars on the fruit does not influence the size of the next season's crop.

It is not often that one can see a population of codlin that is checked by an absolute shortage of food because usually the population is held at a much lower level by repeated poisoning. But recently I came across an orchard in which the codlin had become highly resistant to DDT. Although the orchard had been sprayed 10 times during the summer this was virtually equivalent to not spraying it at all. The codlin multiplied until, towards the end of the season, virtually every apple contained a caterpillar. Because the total stock of food was independent of the size of the population of feeding animals, and because the behaviour of the caterpillars was such that the principle of effective food did not apply, the size of the population of codlin could be predicted by merely counting the number of apples in the area. As with *Cactoblastis*, *Rodolia* and *Ptychomyia* the food for codlin was entirely accessible and the size of the population depended on the absolute amount of food in the area. But codlin presented an exceptionally simple ecological situation because the amount of food was quite independent of the number of eating animals (Andrewartha, 1957).

Small mammals such as mice, voles, lemmings and rabbits produce outbreaks which damage crops but the occurrence of territorial behaviour in these species (sec. 6.122) makes it doubtful whether the idea of absolute shortage of food is useful in explaining their numbers.

7.223 Conservation of wildlife

Territorial behaviour is absent or weakly developed in many ungulates. Consequently they are likely to multiply until food becomes absolutely short unless they are checked by some other component of environment: predators may often be important. Leopold's (1943) account of the deer on the Kaibab plateau in Arizona may be quoted as an example. He considered that the area in its natural condition supported about 4,000 deer. These would have made little impression on their supply of food, and he estimated that in 1918 the area might have supported 30,000 deer without detriment to their stocks of food. From 1907 onward, large numbers of pumas, coyotes and wolves were shot. The deer multiplied rapidly, reaching 100,000 by 1925. A drastic shortage of food during the next two winters resulted in the death of 60,000 deer. At the same time, the feeding of so many animals permanently reduced the total stock of food in the area (Fig. 7.03), and the numbers continued to decline to 10,000 by 1940.

7.3 FURTHER READING

The principles that are outlined in Table 7.01 have been discussed in greater detail by Andrewartha and Browning (1961). This paper also presents the idea of 'absolute shortage of food' in better perspective than I have done in secs. 7.2 *et seq.*

CHAPTER VIII

Components of Environment: a Place in which to Live

8.0 INTRODUCTION

This chapter concludes my analysis of the idea of environment. There are three points that need to be discussed. (*a*) Animals are usually found living in the sorts of places that are proper to their species. And ecologists need to learn the naturalists' art of recognizing the characteristics of these places. (*b*) The influence of weather, predators and other components of environment on an animal's chance to survive and multiply may depend on the sort of place where it is living. And conversely, the degree of protection afforded by a particular sort of place may depend on the sort of weather, predators, etc. that constitute the rest of its environment. (*c*) When there are no other important checks to multiplication the size of the population in a particular area may depend on the number of places that are suitable for the animals to live in.

8.1 THE TENDENCY OF ANIMALS TO CHOOSE A SPECIFIC SORT OF PLACE TO LIVE IN

The passage in E. B. Ford's book *Butterflies* which I quoted in section 1.2, describes how a naturalist, as he becomes familiar with the animals that he studies, comes to recognize the characteristics of the places where each sort of animal is likely to be found.

Glover *et al.* (1955) described how this 'naturalist's art' was used in a campaign to exterminate the tsetse fly *Glossina morsitans* from 280 square miles of bush in northern Rhodesia. It was known that *G. morsitans* rarely hunts for food in dense scrub but only in places where the vegetation is open. So Glover and his colleagues set out to prevent bush-fires and do everything else that they could to make the bush grow more thickly. In this they were largely successful. At the end of 10 years the undergrowth was considerably thicker, but there was not much difference in the density of the taller vegetation. By the end of 5 years the flies (that were taken in a routine sample) had decreased from 5,915 to 4,739 but there was no indication that they were likely to be exterminated from the area by these methods.

About this time it was noticed that *Glossina* tended to be slightly more numerous in certain small valleys which were known locally as 'kasengas'. Glover *et al.* described one characteristic kasenga as follows.

> It starts between high hills as a narrow but flat-bottomed cleft a little under two miles long: after two miles it acquires a definite water-course and becomes a rocky ravine. The upper, flat-bottomed section supports a fair growth of trees, and the adjacent hillsides are also well wooded with *Brachystegia*: the trees in the valley are mainly *Terminalia* and *Combretum*.

On the assumption that the kasengas were important for the tsetse flies, Glover *et al.* decided to clear them. The trees along the small valleys and along the edges of the surrounding woodland were felled. The clearing was started in 1941 and finished in 1944, altogether 2·2% of the total area was cleared.

The results were immediate and spectacular. The number of flies caught in a routine sample which had dropped by about 35% during the 5 years of 'fire exclusion', dropped, during the 5 years after clearing of kasengas began, from 2,750 to 40; by 1950 Glover *et al.* were satisfied that *G. morsitans* had been exterminated from the whole area of 280 square miles.

The final extinction of the population was probably hastened by failure of the sexes to meet and there was some suggestion that right from the beginning the chief outcome of clearing the kasengas was to reduce the likelihood that the flies would mate. Glover *et al.* commented on this as follows.

> In 1942, when fly numbers had already been greatly reduced, an uninseminated though non-teneral female fly was captured in the Isoka area; such an observation had never before been made by the Tanganyika Department of Tsetse Research, who had invariably found that all wild females which had taken their first meal proved on dissection to be inseminated. In the following year, 12 non-teneral females out of 136 (9%) were uninseminated; in 1944 the figure was 20% of 152 and in 1945 it was 29% of 42 dissected. These findings strengthened our belief that there was later no breeding population of flies within the area, since as early as 1942, when numbers were relatively high, the sexes were beginning to fail to encounter each other,

and this effect increased yearly to 1945, after which numbers became extremely low and it seems certain that the great majority of females must then have remained virgin. (See also sec. 6.11.)

It was suggested as a result of these observations that the kasengas might have been important because the flies sought them out and stayed in them until they had mated; but this suggestion was not verified. The authors conclude their paper with the comment: 'It must be confessed that we do not know what vital role these valleys played in the ecology of the flies; we only know that cutting down the trees in and immediately around them put them out of action.'

The ecology of three species of deermice *Peromyscus maniculatus*, *P. californicus* and *P. truei* was studied in an area of 28 square miles near Berkeley, California by McCabe and Blanchard (1950). The area comprised a ridge of hills dissected by many valleys; there was a diversity of aspects and soils supporting woodland, grassland, a dense scrub called 'chaparral' and several other types of vegetation.

The deermice were found only along the margins of the chaparral. None was trapped right inside the chaparral or inside any other type of vegetation. But wherever the chaparral met woodland or grassland, or along the edges of artificial clearings or roads through the chaparral, the mice were found.

The margins differed in their general *facies* and also in their suitability for the different species. The places most favoured by *P. maniculatus* were the moister slopes where a dense stand of chaparral gave way to a narrow band of giant herbs, mostly Umbelliferae with grassland behind them. The ground under the chaparral was bare except for a thin covering of dead leaves.

The other two species were most likely to be found, and their populations were most likely to be dense, along edges of the chaparral where it met woodland of *Pinus*, *Eucalyptus* or *Quercus*. It was characteristic of the places favoured by *P. truei* and *P. californicus* that the chaparral would be composed of a greater variety of species and its margins would be more sinuous than those that were favoured by *P. maniculatus*. Their requirements were more stringent than those of *P. maniculatus* because they were always very few except in places that suited them precisely. But *P. maniculatus* was almost ubiquitous, being present, if not abundant, wherever the chaparral manifested a well-defined edge.

Along some edges that suited *P. maniculatus* well this species made up 95% of the catch. In places that were more suitable for *P. truei* and

P. californicus these species might make up 75% of the catch. No place was found where the population consisted exclusively of one species. So their requirements overlapped to some extent although each species had its own characteristic requirements of the places where it would live.

8.2 SEVERAL EXAMPLES OF HOW THE INFLUENCE OF WEATHER, PREDATORS AND OTHER COMPONENTS OF ENVIRONMENT DEPENDS ON THE SORT OF PLACE WHERE THE ANIMAL IS LIVING

The Colorado potato beetle *Leptinotarsa decemlineata* in Montana burrows into the soil at the end of summer and remains there hibernating during winter. In light, sandy soil they may burrow to a depth of 14 to 24 inches but in harder soils most of them are likely to be within the top 8 inches, concentrated just above the 'plough line'.

Mail and Salt (1933) measured the cold-hardiness of *Leptinotarsa* in the laboratory and concluded that all the beetles would die if they were exposed to −12°C for even a few minutes; more than half would die after several hours at −7°C; but most of them would survive for many days or weeks at −4°C.

They then measured the temperature of the air 6 inches above the ground, and in the soil down to a depth of 6 feet. The measurements were taken twice, once with and once without a cover of snow on the ground. The results are shown in Figure 8.01. In the absence of snow a beetle would have needed to be at least 2 feet below the surface to have had much chance of surviving but in the presence of snow most of the beetles that were 12 inches below the surface would probably have been safe.

These studies showed that the important qualities of the place where *Leptinotarsa* may spend the winter are the nature of the soil (which may determine how deeply they burrow) and the depth and permanence of snow.

The life-cycle of the sheep-tick *Ixodes* was described in sections 6.11 and 7.1. After engorging with blood the tick drops from the sheep and crawls into the mat of vegetation near the ground. It is poor at crawling so it has to seek shelter within a few inches of where it drops. If it is lucky enough to fall into a good place it may survive the 12 months that must elapse before it goes in search of its next meal.

Desiccation is probably the chief hazard for a tick during this period. It can absorb water from the air if the relative humidity is not below about 90% (sec. 5.222); if the air is drier than this the tick loses water by evaporation and has no way of replenishing it. So the only sort of place that gives the tick much chance of surviving during the period between meals is one where the relative humidity of the air is likely to be above 90% for most of the time.

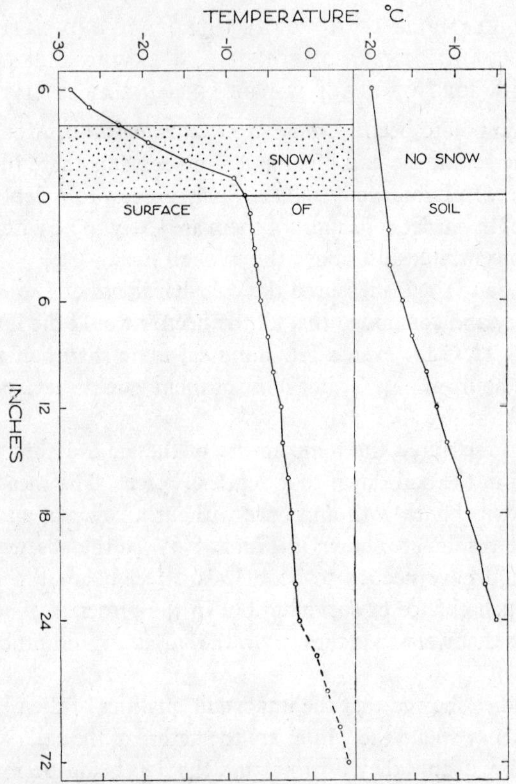

Fig. 8.01. Temperature of the soil at various depths below the surface: *upper curve*, in the absence of snow; *lower curve*, when snow was present. Note the steep gradient through the 4 inches of snow and the consequent less severe temperature below the snow. (After Andrewartha and Birch, 1954 – data from Mail, 1930)

If the mat of vegetation is thick, as it usually is on the lower slopes, it is likely to be moist enough for the tick. If it is thin, as it usually is on the higher slopes and around sheep camps, it is likely to be too dry; the ticks that happen to fall in these places are not likely to survive. The ticks are established only in the rough upland pastures; the finer lowland pastures are too dry and ticks have not become established there although they have repeatedly been introduced.

For many years orchardists near Sydney had to contend with the Mediterranean fruit fly *Ceratitis capitata* which was a serious pest of apricots, peaches and other fruit. At the present time *Ceratitis* cannot be found in this area despite thorough searching; it is almost certainly extinct. Its place has been taken by *Dacus tryoni* which has moved in from the north. This species is now abundant living in almost the same circumstances which previously were characteristic of *C. capitata*. Throughout the whole period fruit has continued to be grown to supply the needs of the million or so people who live in Sydney; so there has at no stage been any shortage of food for fruit flies.

The disappearance of *Ceratitis* has not been explained but at least it seems clear that the fruit which, for many years in the absence of *Dacus*, was a good place for *Ceratitis* to live ceased to be a good place when *Dacus* was present.

Apples in commercial orchards are now almost universally grafted on resistant root-stock which makes the roots an unsuitable place for the aphid *Eriosoma*. The predator *Aphelinus* has been spread to most places where *Eriosoma* occurs. So *Eriosoma* is now a rare species almost everywhere (sec. 6.211). But occasionally one comes across seedling apples, growing on their own roots. In these circumstances *Eriosoma* may flourish on the roots but not on the aerial parts if *Aphelinus* is also present. In the absence of *Aphelinus* both roots and branches would be favourable places for *Eriosoma*.

The European starling was introduced into North America about 1890. It is now widespread and abundant. On the whole it lives in different sorts of places from the bluebird *Sialis sialis* and the flicker *Colaptes auratus* which still have a large part of their populations living away from towns; but neither the bluebird nor the flicker live in towns as much as they used to do before the starling came.

8.3 THE DISTRIBUTION AND ABUNDANCE OF PLACES WHERE ANIMALS MAY LIVE

I mentioned in section 1.1 that outbreaks of the grasshopper *Austroicetes* occasionally do a lot of damage to pastures in South Australia in the area shown in Figure 1.01. Less than 150 years ago, before this area was settled by Europeans, it carried a woodland vegetation of *Eucalyptus* spp. in parts, and in other parts a scrub of *Acacia*, *Kochia*, *Atriplex* and other shrubs. The only places suitable for the breeding of *Austroicetes* would have been grassy glades which occupied only a small proportion of the total area. As the land was occupied the vegetation was cleared. In some areas the land was cultivated for wheat for a number of years and then converted to pasture; in other places pastures were developed from the beginning. But in either case the pastures contained much food for the grasshoppers and many places that were suitable for them to lay eggs in. Several thousand square miles were converted into breeding grounds for *Austroicetes* with a consequent enormous increase in their numbers (Andrewartha, 1943, 1944).

Errington and Hamerstrom (1936) estimated the number of bobwhite quail *Colinus virginianus* that over-wintered in 20 areas in Iowa and Wisconsin during 6 years. They found that each area usually supported a characteristic number of birds. In 4 areas that were studied intensively this varied from one bird to 5 acres to one bird to 65 acres. Errington and Hamerstrom referred to this as the 'carrying capacity' of the area. It seemed likely that the carrying capacity was largely determined by the number of suitable places for quail to live.

The quail form coveys during winter which break up into breeding pairs during summer. The requirements of *Colinus* may be summarized under four headings. (*a*) For nesting, it requires well-drained ground with a thin covering of grass or brush, preferably with some bare ground nearby. (*b*) The nestlings are fed mostly on insects but berries and grains constitute from 80 to 100% of the food of the older birds. (*c*) For protection from bad weather and as a place in which to hide from predators, thickets of scrub or tangles of vines are required, especially during winter. (*d*) For sleeping the birds usually seek a high place. The quail are likely to find a good place to live in an area where meadow, woods, scrub and field meet.

Imagine an area of say one square mile in which these four sorts of country were equally represented but each consisted of solid blocks, occupying one quarter of the area as in Figure 8.02A. It is probable that the central point, where all four areas meet, would form a good place for one covey of quail, but there would be none elsewhere in the square mile. Now imagine the same area of each sort of country arranged patchily as in Figure 8.02B. On this square mile there would be five places where a covey of quail would find all its requirements. The 'carrying capacity' of the second arrangement is higher than that of the first.

The musk-rat *Ondatra zibethicus* usually lives along the edges of shallow water where it builds burrows or 'lodges'. Marshes are good; so are drains if they are fed by tiles, especially if they run near fields that provide food. Streams also provide places for musk-rats to live but according to Errington (1944, 1948) a stream becomes increasingly suitable the more it resembles a marsh. Fluctuations in the level of the water may cause fluctuations in the carrying capacity of the area.

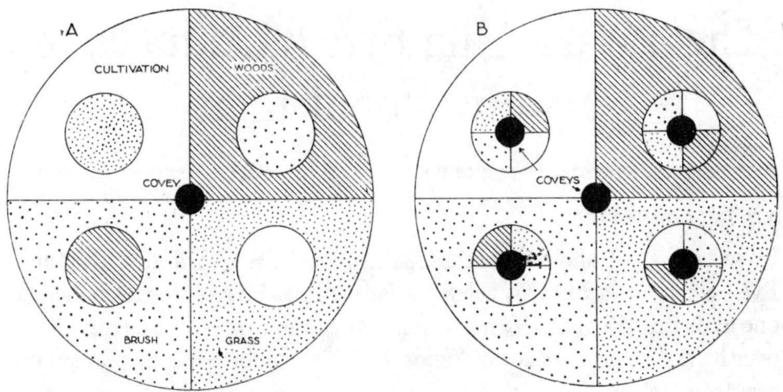

Fig. 8.02. Theoretical distributions of woodland, brushland, meadow and field in two areas to illustrate the meaning of 'carrying capacity' for a species, like the bobwhite quail, which requires all four sorts of vegetation in the place where it lives. (After Andrewartha and Birch, 1954)

Temporary changes in carrying capacity may be brought about by weather. The carrying capacity of large areas has been changed artificially by draining, damming or ditching in areas of marshland or other suitable territory. Dymond (1947) referred to the construction of dams, dikes and canals in the Saskatchewan River, Manitoba, which diverted water into dry marshes and thereby greatly increased the number of suitable places for musk-rats to live: the number of musk-rats in the area increased from 1,000 to 200,000 in 4 years.

8.4 FURTHER READING

Andrewartha and Birch (1954). Chapter 12.
Elton (1949).
Elton and Miller (1954).

CHAPTER IX

Theory: the Numbers of Animals in Natural Populations

9.0 INTRODUCTION

In section 3.3 I showed that animals are usually distributed patchily. Even when one arbitrarily selects what seems to be a uniform small area one finds that the animals are not distributed uniformly or even randomly over it; they tend to occur in 'clumps'. This may be partly explained by accidents in the history of the population, by the behaviour of the animals (e.g. gregariousness) or by the heterogeneity of the terrain. The last-named cause becomes increasingly important as the area becomes large. In large areas, of the order that we usually choose for a practical ecological investigation, irregularities of terrain are pronounced and this, in itself, leads to 'colonial' or 'clumped' distributions of the animals in the area (Elton, 1949). For a description of an extremely colonial distribution see the passages quoted from Cockerell and Flanders in section 4.4. See also section 3.3.

A general theory about the numbers of animals in natural populations, if it is to be realistic, must take into account the uneven distribution of the animals. Andrewartha and Birch (1954) suggested that one way to do this was to regard the whole population as the sum of a number of local populations. They examined the conditions of commonness and rareness in local populations as a first step and then extended the principles to cover the larger population comprising many local populations. I have followed the same method in this book and sections 9.1 and 9.2 are based largely on Chapter 14 of Andrewartha and Birch (1954).

9.01 *The meanings of 'common' and 'rare'*

If we say that a species is *rare* or *common* we imply that individuals of this species are few or many relative to some other quantity that we can measure. We might, for example, relate them to some unit of area and judge them rare if they are, say, fewer than one (or one million) per square mile. This is the commonsense meaning for the hunter and the fisherman, and it is the meaning that I shall use in section 9.2.

Another way of considering commonness and rareness is to relate the number of individuals in the population to the quantities of necessary resources, food, nesting sites, etc., in the area that it inhabits. This may be the commonsense meaning for a farmer who may not be interested in knowing how many caterpillars there are per square mile; it may be more important for him to know whether they are numerous enough to eat much or little of his crop. This is the meaning that I use in section 9.1.

9.1 THE CONDITIONS OF 'COMMONNESS' AND 'RARENESS' IN LOCAL POPULATIONS

9.11 *The conditions of commonness in local populations*

The codlin moth that had become resistant to DDT (section 7.211) were common relative to their stocks of food – so common in fact that there was a caterpillar in almost every apple. But they did not, by virtue of their presence, reduce the size of the next year's crop. In other words, although they multiplied up to the limit of their stock of food they did not, by their numbers, reduce the amount of food for the next generation. A similar condition was described by Kluijver (1951) for a population of great tits *Parus major*. The tits were living in an area of mixed woodland where most of the trees were young with few holes suitable for nests; the birds consistently used up all the nesting sites year after year.

This condition of commonness in a local population may be represented by the curve and the symbols in Figure 9.01. This diagram is generally true of any local population whose numbers are determined by the stock of some non-expendable resource – non-expendable in the sense that the amount available to the next generation is independent of the amount used by the present generation (Table 7.01).

The local populations of *Cactoblastis* on prickly pear in Queensland (secs. 4.5, 7.221) represent the converse of this condition. The first ones to arrive at a place where prickly pear is growing multiply rapidly. But the food, being a growing plant, tends to be reduced in relation to the number of *Cactoblastis* feeding on it (sec. 7.21); so there is less food for each succeeding generation; eventually there comes a time when there is none left for the next generation; and the population in that locality dies out – but not, as a rule, before it has given rise to emigrants, some of which may find a fresh supply of prickly pear and start the cycle all over again in a new place. This sequence of events is illustrated by the diagram in Figure 9.02. This diagram is generally true of all local populations whose numbers depend on the amount of some diminishing or expendable resource – expendable in the sense that the more that is used by one generation the less there will be for the next.

9.12 *The conditions of rareness in local populations*

The woolly aphis which I discussed in section 6.211 never consumes more than a fraction of the food in its area. During the summer no local colony of *Eriosoma* lasts for long without being discovered by the predator *Aphelinus*; and shortly after it has been discovered it is annihilated – but usually not before a few emigrants have gone off with a chance of founding a new colony on a neighbouring tree. Thus the aphid remains rare relative to the large amounts of easily accessible food that surround it (see also sec. 4.4). I have illustrated this sequence of events in Figure 9.03 with the series of symbols that ends with a circle that is empty except for the cross inside it; this symbol indicates that there used to be a colony of animals here until recently.

As winter approaches, the lower temperatures at first hamper the development of *Aphelinus* relatively more than that of its prey. Colonies of *Eriosoma* that are discovered by *Aphelinus* late in summer may not be annihilated before both species become dormant for the winter. With rising temperatures in the spring *Eriosoma* is able to multiply while *Aphelinus* still remains dormant. This sequence of events is represented in Figure 9.03 by the series that ends with a circle with a small segment shaded; this symbol indicates that a small remnant of the population persists and becomes the nucleus which multiplies again when circumstances permit. The overall size of the circles remains unchanged because the animals are few relative to the amount of food; so they make no appreciable difference to the amount that is present for succeeding generations.

The diagram in Figure 9.03 with its two alternative endings has a wide application. It can be taken quite generally to represent the broad and important condition of rareness in which any component of environment keeps the population rare relative to its stock of food, nesting sites, etc. I have discussed this diagram in relation to a population that was kept in check by a predator, but the check might equally well have been exercised by weather, the recurrent use of an insecticide or any other component of environment.

I mentioned in section 8.2 that when *Eriosoma* colonizes a seedling apple that is growing on its own roots the aphis can thrive on the roots free from attack by *Aphelinus*. But those that colonize the branches are almost certain to be destroyed by the predator quite soon. I have represented such a situation in Figure 9.04. This diagram may be taken generally to represent any population in which a more or less constant proportion is secure against a hazard that is likely to destroy the remainder. It is a situation that one does not meet very often among invertebrates but it represents all those vertebrates that manifest territorial behaviour (sec. 6.122).

9.13 *The way in which weather may keep a local population rare relative to food and other resources*

In regions with temperate climate, and perhaps in most other climates as well, many species have well defined breeding seasons and most species experience only a limited period that permits multiplication. For the rest of the year deaths exceed births and the numbers decrease. Consequently it is characteristic of populations whose numbers are determined largely by weather, that they should fluctuate more or less in step with the seasons (Fig. 7.01). And what is true of seasons may also be true of longer periods. Runs of favourable years may be followed by runs of less favourable ones. Thus the average density that is attained over any period of years depends on the relative rate of increase and decrease during the favourable and unfavourable periods, and the relative duration of the favourable and unfavourable periods. I have illustrated this principle in Figure 9.05 (see also Fig. 1.03). In order to simplify the exposition I have drawn three pairs of curves so that rate of increase, extent of decrease and duration of favourable period may be compared independently. But in nature these three comparisons may not be independent of each other. The difference between two places at the same time or the same place on separate occasions may be compounded of all three comparisons.

In Figure 9.05 the abscissae are in units of time – days, years or generations. The ordinates for each curve are numbers of animals; K represents the number that the area would support if the total resources of food were used up. The number indicated by Y_0 is the number left when the unfavourable period ends and is also the nucleus for multiplication when the next favourable period begins. The rate of increase that pertains during the favourable period is called r. Suppose that the three pairs of curves represent three ways that weather may influence the numbers of animals in a local population. The curves on the left relate to a place where the weather is favourable, and those on the right relate to a place where the weather is severe. In the top pair the two curves start from the same level at the beginning of the favourable period (Y_0 is constant); and they rise at the same rate because r is constant. But the curve for area A rises farther because the favourable period lasts longer. With the middle pair the two curves rise at the same rate (r is constant); they continue rising for the same time (favourable period, t is constant); but the curve for area A rises farther than that for B because it started from a higher level (Y_0 is greater for A). This means that for some reason or another the unfavourable period was less prolonged or less severe in A than in B so that the population in A was still relatively large when the weather changed and allowed the animals to start increasing again. With the bottom pair the two curves start from the same level (Y_0 is constant); they continue rising for the same interval of time (favourable period, t, is constant); but the curve in A rises farther because it is steeper (rate of increase, r, is greater for A). Taking all three curves into account and compounding them, one can easily see that the animals would, on the average, be more numerous in area A than in B. The same idea is expressed differently in Figures 1.02 and 1.03.

Figure 9.05 may also be considered to represent the condition of rareness that prevails in a population of insects that is 'controlled' by insecticides. The first pair of curves compares the results of frequent and infrequent applications of the insecticides. The second pair of curves compares the result of spraying thoroughly with a good insecticide with the result of spraying less thoroughly or with an inferior insecticide. The third pair of curves illustrate what might happen if the same treatment were used against two populations of the pest one of which was able to multiply more rapidly than the other.

9.2 THE CONDITIONS OF 'COMMONNESS' OR 'RARENESS' IN NATURAL POPULATIONS

When we come to consider the whole population in an area that is large enough to support many local populations it is necessary to broaden the meaning of 'common' and 'rare' to include the idea of so many per unit area (section 9.01 – first meaning). There are four points to be considered. (*a*) What are the conditions of commonness or rareness in the local populations? (*b*) Are the local populations in step? Or are different ones at different stages of development at the same time – some increasing while others are decreasing and so on? (*c*) How many local populations are there in a unit of area and how large are the individual local populations? (*d*) How relevant is the principle of relative shortage of food (sec. 7.1)?

With these four points in mind I have constructed a series of diagrams which are intended to illustrate the conditions of commonness or rareness in natural populations, i.e. populations that occupy substantial areas of the dimensions that ecologists usually choose to study.

In Figure 9.06 each symbol represents the condition, at one moment of time, in a local population of the sort that can be represented by Figure 9.01. Area A may be compared with area B. In both areas the local populations are similar (in that the animals are common relative to the stock of some unexpendable resource) but there are more of them in A than in B. Consequently the population in the whole area is more numerous in A than in B.

In Figure 9.07 each symbol represents the condition, at one moment of time, of a local population of the sort that can be represented by Figure 9.02. It is characteristic of this condition that the local populations are likely to be at different stages of development at the same time (see the account of *Cactoblastis*, secs. 4.5, 7.221), and I have allowed for this in drawing the diagram. I have also allowed for differences in the frequency of local accumulations of food and in the dispersiveness and ability to search, of the different sorts of feeding animals. There are, therefore, three comparisons that may be made within Figure 9.07. Both the species that occur in area A and the one that is found in areas B and C multiply, in local populations, to the limit of the food supply and then die out from an absolute shortage of food (sec. 7.2). The species in area A is highly dispersive; the one in areas B and C is less so. In considering the whole area it would seem clear that the animals in C would experience a severe shortage of food in the relative sense; and in a different way the same is true of those living in A also. The animals in A are so dispersive that very little food goes undiscovered for very long; consequently the food becomes scarce and sparsely distributed, and the animals come to experience a severe relative shortage of food.

Figure 9.08 is made up from symbols taken from Figure 9.03 in order to illustrate the way that a predator may influence the abundance of its prey. The condition of the population in A may be compared with the description of *Saissetia* (as prey for *Rhizobius* in California) and the condition of the population in area B may be compared with the description of *Icerya* in section 6.211.

The two areas A and B support populations of the same species of herbivore. These herbivores are not numerous enough to reduce their stocks of food appreciably, so there are the same number of circles drawn in both areas. In area B there is an active predator with powers of dispersal and rate of increase which match those of the prey; in area A these predators are replaced by another species with equivalent rate of increase but with inferior dispersiveness. (That is, it annihilates local populations just as effectively but it finds them less frequently.) In B many localities that are quite favourable are empty; some have become empty only recently (circle with a cross in it), and in most of those that are occupied the numbers are low. Either the prey have only recently arrived at the place and have not yet had time to become numerous, or else they have been there longer and are in the process of being exterminated by the predators. Contrast this with A where relatively more places are occupied and the local populations are, on the whole, larger. There are relatively few places from which the prey have recently been exterminated. The prey in area A have the advantage of being more dispersive than their predators and this allows them a longer interval for multiplication in each new place that they colonize.

The way that weather influences the density of the population may be illustrated in Figure 9.09. It is based on Figure 9.05. In Figure 9.09 there are the same number of places where the animals may live in A and B, but the weather is more favourable in A than B, so the animals, in each local population, are more common, relative to their stocks of food, etc., in A than in B; consequently the population in the whole area is larger in A than in B. Area C is inferior to B with respect to the number of places and C is inferior to A with respect to both the number of places and the weather. So the animals are rare in C relative to A, in both meanings of the word (sec. 9.01).

9.21 *General conclusions*

Figures 9.03, 9.04, 9.05B, 9.07B, C, 9.08B and 9.09B, C, illustrate populations in which animals are rare relative to their stock of food (or other resource). This is true of the whole population in Figure 9.07C even though the local populations consume all their food. In Figure 9.07 (not in the local populations but in the area considered as a whole) the number of animals is determined chiefly by a relative shortage of food (sec. 7.1). In every other instance where the animals are rare relative to their stock of food (or other resource) the number in the population is determined by a shortage of time. This is immediately obvious from inspection of Figure 9.05. But inspection of Figures 9.03 and 9.08 also leads to the same conclusion: each local population is likely to be exterminated soon after it has been found by the predators; how numerous the animals become in each local population depends largely on the interval of time that elapses between the founding of the colony and its discovery by the predators; when this interval is long the animals become relatively abundant, when it is short they remain relatively scarce.

Figures 9.01, 9.02, 9.05A, 9.06, 9.07A, 9.08A and 9.09A illustrate populations in which the animals are abundant relative to their stock of food (or other resource). When, as in Figures 9.01 and 9.06, the resource is not expendable, the number in the population depends quite simply on the total amount of resource. When the resource is expendable, as in Figures 9.02 and 9.07, the number in the population may still be said to depend on the amount of resource but the relationship is more subtle because the amount of resource also depends on the number of animals: that is, the relationship is to some extent reciprocal. In the local populations the numbers can be explained in terms of the absolute shortage of food (sec. 7.21; Fig. 7.03); in the natural population, comprising many local populations, the number of animals depends on the absolute amount of resource and its accessibility (shortage in the relative sense, sec. 7.1).

Thus we conclude that the density of natural populations may be explained in terms of:

(i) a shortage of time,
(ii) a relative shortage of some essential resource,
(iii) an absolute shortage of some essential resource, or interactions between these three mechanisms.

I have arranged them in the order of what I judge to be their relative importance in nature. I would like to emphasize that this statement applies to the *number of animals that have been counted in a natural population* because, as I said in sections 1.1 and 3.0, this is the basic material that the ecologist has to work with.

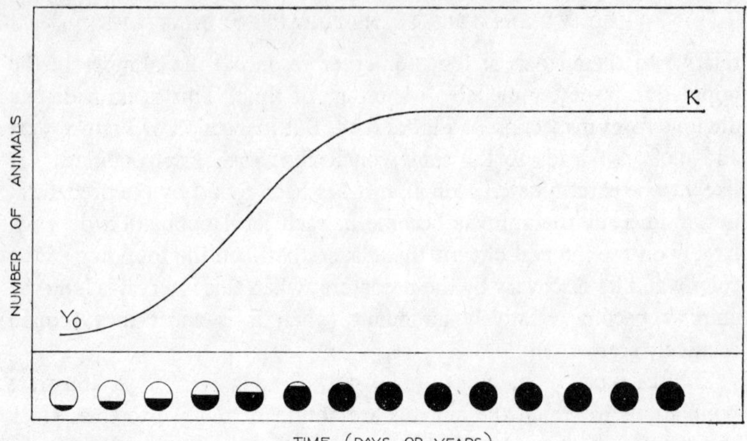

Fig. 9.01. The growth of a 'local population' (i.e. the population in a 'locality') whose numbers are limited by the stock of some non-expendable resource, such as nesting sites. The initial numbers (i.e. the number of immigrants who originally colonized the locality) are represented by Y_0; the maximal numbers that the resources of the locality will support are represented by K. The circles repeat the information given by the curve. The proportion of the circle shaded represents the number at a specific time as a proportion of the maximum.

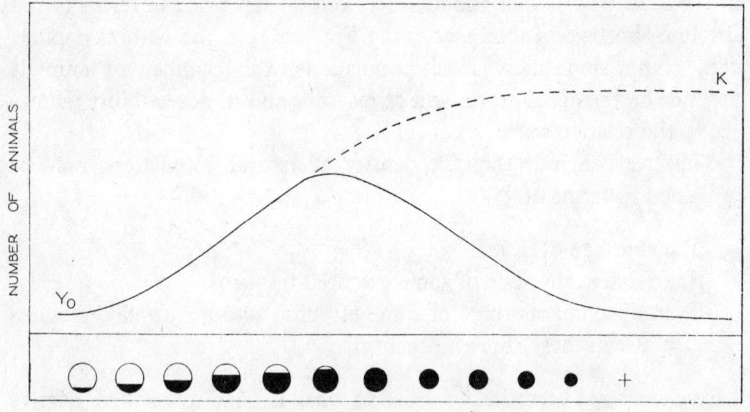

Fig. 9.02. The solid line represents the growth of a local population whose numbers are limited by a diminishing resource, such as food, which becomes less, the more animals there are feeding on it. (The broken line indicates the numbers that the population might attain if the resource were non-expendable, as in Fig. 9.01.) The symbols have the same meaning as in Fig. 9.01; the circles grow smaller because the plants (or population of prey) become fewer and eventually die out because there are too many animals eating them. The cross indicates that this is a place where a local population recently became extinct (see Figs. 9.07 and 9.08). (After Andrewartha and Birch, 1954)

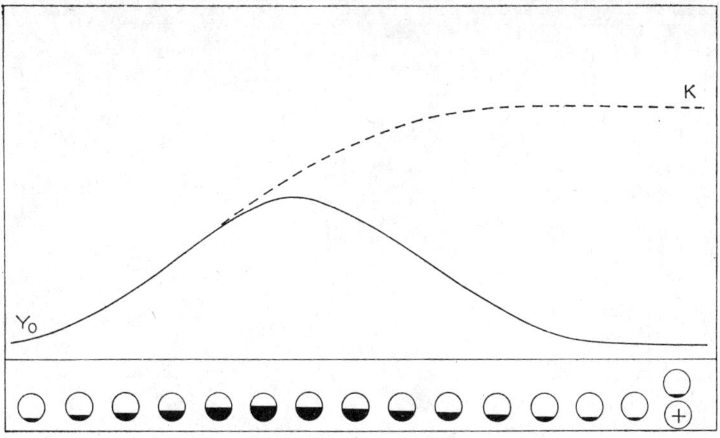

Fig. 9.03. The solid line represents the growth of a local population that never uses up all its resources of food, etc., in the locality because its numbers are kept by weather, predators, shortage of food in the relative sense (sec. 7.1), or some other environmental component, at a level well below that which the resources of the locality could support. The broken line and the symbols have the same meaning as in Figures 9.01 and 9.02. The two symbols at the end indicate alternative endings to the cycle of growth of the local population: the population may be extinguished in this locality (*circle with a cross in it*) or a remnant may persist, perhaps to increase again when the circumstances change (*circle with a remnant of shading in it*). The circles do not become smaller because the resource, although it is of the sort that would diminish if many animals used it, nevertheless does not diminish appreciably because the animals do not become numerous enough (sec. 9.32). (After Andrewartha and Birch, 1954)

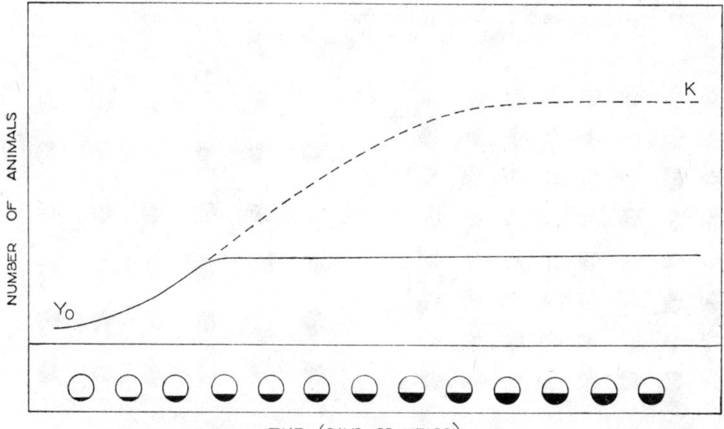

Fig. 9.04. The solid curve represents the growth of a local population in which a more or less constant number of the individuals are sheltered from some hazard which is likely to destroy all those that are not so sheltered. This number is not enough to use up more than a small proportion of the resources of food, etc., in the locality. (After Andrewartha and Birch, 1954)

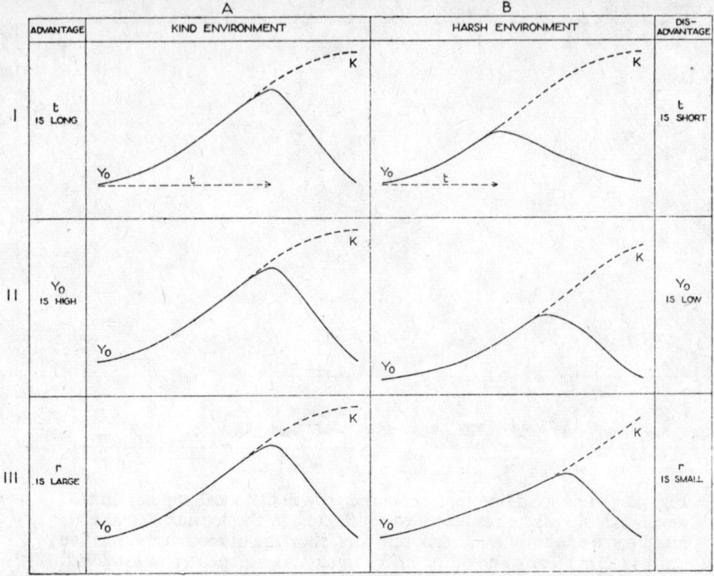

Fig. 9.05. Three ways in which weather may influence the average numbers of animals in a locality. The numbers in the locality are increasing during spells of favourable weather and decreasing during spells of unfavourable weather. Two areas A and B are compared with respect to three qualities. Quality I is related to the duration of the favourable period; quality II is related to the severity of the unfavourable period; and quality III is related to the favourableness of the favourable period. The numbers that would be attained if all the resources of food, etc., were made use of are indicated by K; the numbers to which the population declines during the unfavourable period are represented by Y_0. For further explanation see text. (After Andrewartha and Birch, 1954)

Fig. 9.06. The populations of large areas are made up of local populations. In the two areas which are compared in this diagram the resources of every locality are fully used up (as in Fig. 9.01) but there are more favourable localities in area A than in area B. (After Andrewartha and Birch, 1954)

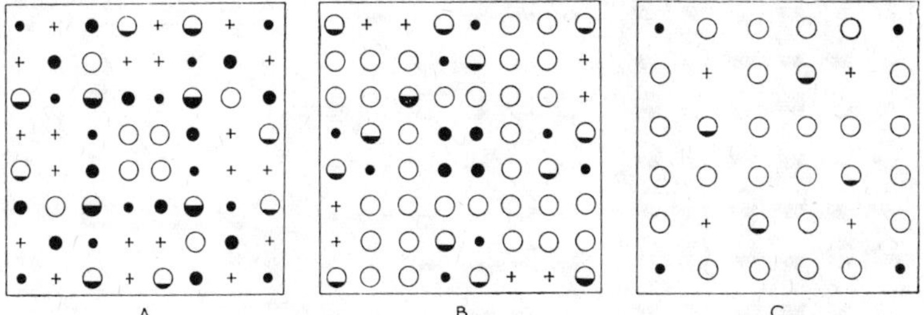

Fig. 9.07. The populations of the large areas A, B and C are made up of a number of local populations. The symbols describing the local populations are taken from Figure 9.02. Favourable localities are distributed equally densely in A and B but more sparsely in C. Area A is occupied by a species with high powers of dispersal; B and C are occupied by a species which is similar except that it has inferior powers of dispersal. In area A shortage of food in the absolute sense may be an important cause of death and the number of animals in the area may be largely explained in terms of this idea. In areas B and C shortage of food in the relative sense (sec. 7.1) is likely to be more important. The number of animals in B is likely to be greater than in C because favourable localities are more densely distributed in B than in C (Fig. 8.02). (After Andrewartha and Birch, 1954)

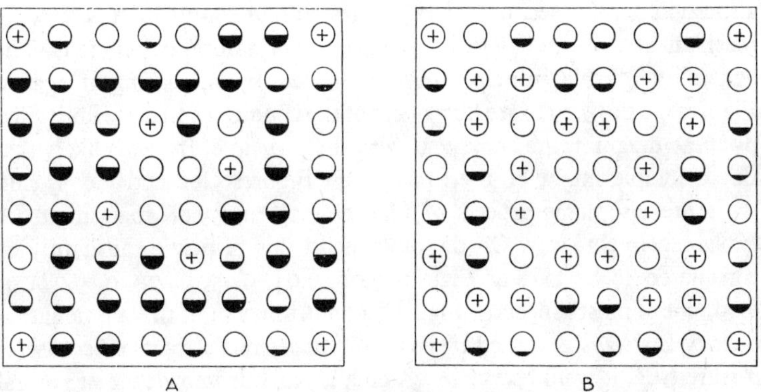

Fig. 9.08. The populations in the large areas A and B are made up of a number of local populations. The symbols describing the local populations are taken from Fig. 9.03. Favourable localities are distributed equally in the two areas. In area B there is an active predator whose powers of dispersal and multiplication match those of the prey. This predator is absent from A where its place is taken by one which is more sluggish. Relatively more of the favourable places are occupied in A than in B, and the local populations are, on the average, larger. Altogether the animal is more abundant in A than in B. (After Andrewartha and Birch, 1954)

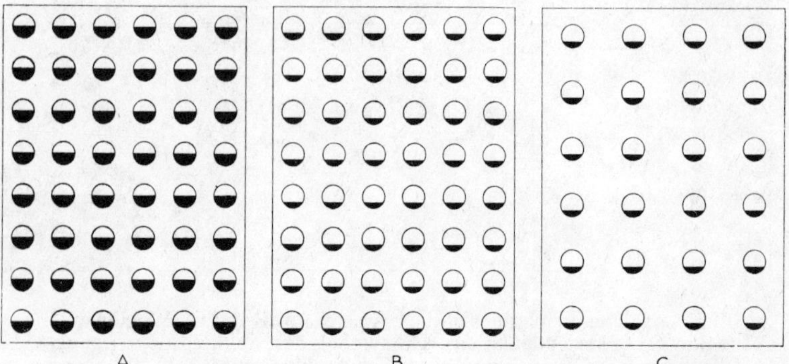

Fig. 9.09. The populations in the large areas A, B and C are made up of a number of local populations. The climate in A is more favourable than that in B or C. The number of favourable localities is the same in areas A and B but fewer in C. For further explanation see text. (After Andrewartha and Birch, 1954)

9.211 Competition

An alternative principle was put forward by Nicholson (1933, 1954; see also quotations in sec. 9.31) who suggested that competition is the only mechanism that can determine the density of natural populations. The generality of this principle may be tested by measuring it against Figures 9.01 to 9.09 to see how many of the conditions that are illustrated by these diagrams are consistent with the principle. It seems likely that this principle might be used to explain Figures 9.02 and 9.07A; and there are also some aspects of Figures 9.03 and 9.08 that might be explained in this way. On the other hand this principle can hardly be considered to be true of Figures 9.01, 9.04, 9.05, 9.07B, C, 9.08A or 9.09, nor of most aspects of Figure 9.08B. At first sight the statment that density is always governed by competition seemed attractive because, if it were true, it would serve to systematize all our knowledge of natural populations in one comprehensive generalization like the famous generalizations of physics. Unfortunately it is not true of natural populations except, perhaps, for a few relatively unimportant conditions of commonness (secs. 6.11, 9.32).

There are two other statements about competition that are referred to fairly often in the ecological literature. They are (*a*) that competition determines the limits to the density of a population (Nicholson, 1933; Elton, 1949) and (*b*) that competition is the only mechanism that can keep a population 'in being', i.e. from becoming extinct (Milne, 1957;

Nicholson, 1954, 1958; see also quotations in sec. 9.31). These s ments can scarcely be intended to relate to the numbers of real an that one may count in a natural population because, as I have pointed out elsewhere (Andrewartha, 1957, 1959), 'ultimate limits' and 'causes for extinction' cannot be inferred from the results of experiments or observations but have to be discovered by deduction or intuition. To criticize these ideas one must consider what part deduction and intuition play in the discovery of scientific knowledge. I discuss this question in section 9.3. The reader who does not wish to follow this theoretical controversy may conveniently stop his reading of Part I at this point. But the student who is going to work through the experiments in Part II should read section 9.33 as an introduction to the practical course.

9.3 THE CONTROVERSY ABOUT 'DENSITY-DEPENDENT FACTORS' AND COMPETITION

The 'theory' of 'density-dependent factors' has been the subject of a controversy in recent years. I shall suggest in section 9.34 that it should not be regarded as a scientific theory at all because its authors have underrated the empirical part of scientific method. This 'theory' is indeed more like dogma because it rests primarily on insight and deductive logic.

9.31 *The idea of density-dependent factors*

So far as I can discover the idea of density-dependent factors was first mentioned in the scientific literature by two American entomologists, Howard and Fiske (1911), who defined 'catastrophic mortality factors' as those that killed a *constant proportion* of the population independent of its density, and 'facultative mortality factors' as those that killed an *increasing* proportion of the population as the density increased. Howard and Fiske did not mention that there are logically two ways in which the 'facultative factors' might produce the result that was ascribed to them. (*a*) Suppose the facultative factor to be a species of obligate predator such that it has no source of food other than a particular species of prey. Suppose also that any increase in the population of prey resulted in a relatively greater increase in the population of predators. Thus we might say that the 'intensity' of the factor had increased because the number of predators had increased. This often happens among insects. (*b*) Suppose the facultative factor to be a species of predator that has many other sources of food and no great preference for eating the particular species that we are considering. It is likely that the predators would eat an increasing proportion of the population as it became denser, but because the predators were merely substituting an easily found food for a more difficult one this is not likely to cause an appreciable increase in the population of predators. Thus we might recognize an increase in the 'efficiency' of the factor because the same number of predators were eating a greater proportion of the prey. This often happens when the predator is a vertebrate.

In both instances the predators comply with the definition of 'facultative factor' but in the first instance the essential relationship of density-dependence is associated with changes in the *intensity* of the factor whereas in the latter the relationship is associated with changes in the *effectiveness* of the factor without any change in its intensity. Smith (1935) also overlooked this distinction. He merely took over the idea of Howard and Fiske without changing it very much, and called the two sorts of 'factors' which they had described density-independent and density-dependent. Nicholson (1933) restricted the term 'density-dependent' to those 'factors' that varied in intensity. Varley (1947) recognized the difference between density-dependent factors that varied in intensity and those that did not; and he called the former 'delayed density-dependent factors'. The position was greatly complicated by Nicholson (1954) who described 9 different sorts of factors and invented new and allegorical names for them. Andrewartha (1957) constructed a key like those used in taxonomy which showed the logical relationships between the different sorts of factors that had been described by Nicholson and others. But there is no need to discuss the classification of factors in detail. It is sufficient to understand the primary distinction between density-dependent and density-independent factors.

Howard and Fiske assumed, without publishing the evidence on which they based their assumption, that weather and other 'physical factors' always act independently of the density of the population, whereas the influence of predators and other 'biotic factors' usually depends on the density of the population. But when I (Andrewartha, 1957) examined a number of 'case-histories' from the ecological literature I found that this assumption by no means corresponded to the realities of nature. Quite often predators and food, which are clearly 'biotic', had to be classified as density-independent, whereas weather, which is clearly 'physical', sometimes had to be classified as density-dependent (sec. 5.0).

The idea of density-dependent factors has been accepted quite uncritically and repeated many times since it was first published by

Howard and Fiske. It appears in different forms in many scientific textbooks and journals. Moreover a number of authors have used this idea as a premise on which to build a more complicated argument. In broad outline, and greatly simplified, the argument runs as follows. A population that is, on the average, increasing must continue to increase indefinitely unless it is controlled by a factor that increases its depressive influence as the density of the population increases. Conversely a population that is, on the average, decreasing must continue to decrease indefinitely (i.e. become extinct) unless it is controlled by a factor that decreases its depressive influence as the density of the population decreases. Because the only density-dependent factors are biotic factors it follows that biotic factors are the only sort that can control a population. And a corollary to the main argument states that because 'biotic factors' always refer to the relationships between animals and their food, their predators or other organisms, therefore 'biotic factors' can always be regarded as some form of competition either within species or between species or between predators competing for their prey. Therefore 'competition' is the only 'factor' that can 'control' a population.

Milne (1957) was quite explicit with respect to the meaning of 'control': a population that is 'under control' is simply one that has not yet become extinct. Thus by definition, any population that is still extant is 'under control'. Williamson (1957) was also quite explicit in accepting this meaning for 'control'. But Nicholson is ambiguous. In some passages (see quotations below) he seems to argue, like Milne, that it is competition that keeps a population 'in being' except that he seems to use 'govern', 'regulate' and 'control' synonymously. In other passages 'control' (or one of its synonyms) seems to relate to the actual numbers that can be counted in a natural population during the period of a particular investigation and especially to fluctuations away from some hypothetical 'equilibrium density'. In so far as these statements are intended to *summarize empirical facts* about the numbers of the real animals that have been counted in a natural population they may be measured against the criticisms of sections 7.2, 9.211 (first paragraph), 9.32 and 9.34. In so far as the statements depend on intuition or logical deduction they may be measured against the criticism in section 9.33. Rather than paraphrase the exponents of this difficult and controversial idea it seems best to let them speak for themselves. So I have selected seven short passages which represent the arguments that have been advanced in support of the 'theory' of competition.

'For the production of balance it is essential that a controlling factor should act more severely against an average individual when the density of animals is high, and less severely when the density is low. In other words, *the action of the controlling factor must be governed by the density of the population controlled.* Clearly no variation in the density of a population can modify the intensity of the sun, or the severity of frost, or of any other climatic factor . . . [But see sec. 5.0.] A moment's reflection will show that any factor having the necessary property for the control of populations must be some form of competition.' (Nicholson, 1933, pp. 133 and 135.) [The emphasis is on imagination and insight throughout this and the following passages, especially in the wording 'A moment's reflection will show . . .']

'Logical argument based upon certain irrefutable facts shows, not merely that populations may regulate themselves by density-induced resistance to multiplication, but that this mechanism is essential. [The logical argument follows together with a reference to Nicholson, 1954, p. 59.]

'Thus it seems impossible to conceive of any situation in which populations do not adjust their densities in relation to the environmental conditions by themselves inducing a change in resistance to multiplication, when their densities are inappropriate to the prevailing conditions. This is the basic justification for the *belief* that all *persistent* populations exist in a state of balance in their environments; for "balance" refers to such a condition of corrective reaction to change which holds a system *in being*.' (Nicholson, 1958, pp. 114, 115.) [My emphasis. I have emphasized '*persistent*' and '*in being*' because this is one passage where Nicholson seems to have accepted Milne's definition of 'control'; I have emphasized '*belief*' in order to make the point that this and nearly all of Nicholson's writings on this subject, properly have the status of what I have called 'a conceptual model'. (See below.)]

'To account for the plagues [of *Euxoa*] we need to know more than this. . . . We need to know what controls the density and not merely what factors may modify it. Exactly what factor controls the population density, it is impossible to say, but there can be little doubt that it is some form of competition.' (Nicholson, 1937, p. 106.) [In this passage 'control' seems to mean 'determine'; there is no suggestion of the meaning in which Milne uses 'control'.]

'Populations are self-governing systems. They regulate their densities in relation to their own properties and those of their environments. . . . The mechanism of density governance is almost always intraspecific competition, either amongst the animals for a critically important requisite, or amongst natural enemies for which the animals concerned are requisites.' (Nicholson, 1954, p. 10.)

'It follows directly from these considerations that populations should exist in a state of balance in all favourable environments and that . . . density-controlled compensatory reaction enables populations to adjust themselves to severe external stresses and *so to remain in being*.' (Nicholson, 1954, p. 60.) [My emphasis.]

'Like the weather, this [relative food shortage] may have an important influence on the rates of mortality and reproduction, but in the same way at high and low densities of the population. Hence it cannot exercise natural control unless its action is linked in some way with a density-dependent process, e.g. by crowding effects between thrips on the flowers. . . . (Solomon, 1957, p. 132.)

'It is difficult to see what factors may be responsible for regulating populations of *Glossina*. Unfavourable conditions of climate are presumably independent of the density and cannot regulate it. Competition for blood between individual tse-tse flies does not, we believe, take place. We are left to suppose that the "enemies" (using that word broadly) of the fly or of the puparium must be responsible for the regulation. But it must be admitted that we have no evidence that any particular enemy becomes more numerous or effective at higher densities of the tse-tse population. We have indeed very little knowledge of the causes of mechanisms which prevent an indefinite increase of these insects when external conditions are favourable, but one must suppose that some such factor exists.' (Buxton, 1955, p. 486.)

There are four epistemological errors represented in this group of quotations. (*a*) These statements have not been tested empirically (i.e. by experiment and observation) and yet their authors have put them forward as if they were established facts about nature (sec. 9.33). (*b*) The use of 'competition' in this context is misleading or unnecessary (sec. 9.32). (*c*) The assumption that biotic factors are equivalent to density-dependent factors is not true to nature (Andrewartha, 1957). (*d*) There is a particular contradiction that becomes apparent when one considers the relationship between the ideas of distribution and abundance (sec. 1.1). If, as Nicholson sometimes seems to claim, abundance must be regulated by density-dependent factors how can one explain the gradual decline in abundance towards the margin of the distribution? And how can one explain the decline to zero at the margin of the distribution? Nicholson does not seem to arrive at an adequate solution to this problem. In one place (1947, p. 11) he said that density-independent factors 'produce either tolerable or intolerable conditions'. If they are intolerable the species cannot exist, if tolerable the animals multiply indefinitely unless checked by density-dependent factors. Yet elsewhere (1958, p. 108) he wrote:

> When there is a gradient from inherently favourable to unfavourable conditions, the mechanism described [density-dependent factors] permits the species to persist indefinitely in all favourable places; and it causes the intensity of induced resistance to fall towards zero as the limits of unfavourability are approached. Consequently, near the fringes of the distribution, density-governing reaction [i.e. density-dependent factors] must be slight and inconspicuous.

9.32 *Competition*

It is difficult to discuss the idea of competition critically because ecologists have made this word into what Hardin (1956) called a 'panchreston'. By analogy with a 'panacea', which is a remedy that cures every ill, a panchreston is a word that explains everything. Hardin wrote: 'The history of science is littered with the carcasses of discarded panchrestons: the Galaenic *humours*, the Bergsonian *elan vital*, and the Drieschian *entelechy* are a few biological examples in point. A panchreston which "explains" all *explains* nothing.' I hope that we ecologists may soon be able to add 'competition' to this list of 'abandoned carcasses'.

Birch (1957) searched the ecological literature and found four meanings for competition, some of which, he said, were quite misleading. He decided that the following meaning was the only one that was not better conveyed by some other word already in common speech. 'Competition occurs when a number of animals (of the same or of different species) utilize common resources the supply of which is short; or, if the resources are not in short supply, competition occurs when the animals seeking that resource nevertheless harm one another in the process.' I would go further than this and restrict the use of 'competition' to the first meaning. My reasons for excluding the second meaning are given towards the end of this section.

Andrewartha and Birch (1954, p. 20) discussed the usefulness of the idea of competition. What follows has been taken largely from that discussion. It seems pertinent to ask two questions. (i) How commonplace is competition in nature? (ii) When competition does occur how much does the idea of competition help us to explain the numbers of animals that we have counted in the population? Accepting Birch's definition we can resolve the first question into two. (*a*) How usual is it for the animals in a natural population to run short of food or some other essential resource? (*b*) How often in nature do we find animals harming each other in their quest for food or some other resource that they share when it is not in short supply?

The writings of almost every ecologist who has discussed the matter make it clear that there must be very few species of animals that are likely ever to experience a widespread shortage of food or any other essential resource. Darwin in Chapter II of *The Origin of Species* wrote: 'Rarity is the attribute of a vast number of species of all classes, in all countries. If we ask ourselves why this or that species is rare, we answer that something is unfavourable in its conditions of life; but what that something is we can hardly ever tell.' Elton (1938, p. 130) wrote: '. . . animal numbers seldom grow to the ultimate limit set by food supply and not often (except in some parts of the sea) to the limits of available space. This conclusion is also supported by the general experience of naturalists that mass starvation of herbivorous animals is a comparatively rare event in nature, although it does occasionally happen as with certain caterpillars that abound on oak trees in some years and may cause complete defoliation. With predatory animals it probably happens more often.' Smith (1935, p. 880) wrote: 'In general, attempts to determine abundance or scarcity have been based upon studies of species which are of economic importance only. The fact that the number of species which become sufficiently abundant to damage crops is relatively small, and that such species form only an insignificant fraction of the total number of phytophagous insects is ignored.' A graphic picture of the rareness of certain species of scale insects relative to their food is given by Cockerell (1934), in the passage that I quoted in section 4.4. And, in general, the evidence would seem to suggest that competition, in this meaning of the word, is rather unusual in nature.

But according to the broader meaning that is also included in Birch's definition, we must call it competition when one animal harms another as both seek the same resource even though it is not in short supply. When two species of blowfly *Lucilia sericata* and *Chrysomyia albiceps* were reared together in a piece of carrion many more of the larvae of *L. sericata* died than when they had been reared equally crowded but without any *C. albiceps* in the carrion. They did not die from shortage of food but because the larvae of *C. albiceps* attacked and destroyed them. It would be misleading to consider *C. albiceps* as a predator because they thrive in carrion without any *L. sericata* in it; and *L. sericata* is certainly not a usual or a necessary part of their diet. Ullyett (1950) called this relationship competition and Birch accepted it as coming within the scope of his definition. Brian (1952) observed that the ant *Formica* sp. in a cut-over pine wood attacked and destroyed ants of another species, *Myrmica* sp. Humming birds of the species *Callype anna* may chase those of another species, *Selasphorus sasin*, from their territories by 'aggressive displays' (Pitelka, 1951). By searching the literature you would find other examples of this sort of 'competition' (Elton, 1958). But I do not know of any review that has assessed its commonness or rareness in nature. In my own experience, I have come across it in relatively few of the species that I know or have read about and I would judge it to be rather unusual. So the answer to the first question that I have asked above would seem to be that no sort of competition is very commonplace in nature. Nevertheless we may occasionally come across competition and we would like to know whether, in these particular instances, the idea of competition helps us to explain the distribution and the abundance of the species. The best way to think about this question is to consider a simple situation in which competition undoubtedly is occurring. For example, Andrewartha and Birch (1954, p. 23) described an imaginary population of solitary bees which we purposely kept simple in order to emphasize the competition.

> Consider a population of solitary bees living in an area in which there are several fencing posts. In the posts there are 500 auger-holes, each one of which will serve as a nest for one bee and one bee only, and there are no other places in this area that are suitable for nests. The bees are skilful searchers and no auger-hole remains unused while there is still a bee requiring a nesting site. Each bee needs to make only one nest during her lifetime. Now on one occasion we shall suppose that there were 500 bees present; they filled up the 500 auger-holes and no competition occurred. On a second occasion we shall suppose that there were 1,000 bees; they also filled up the 500 auger-holes, but only after a certain amount of competition that resulted in 50% of the population being deprived of the chance of

contributing progeny to the next generation. On a third occasion we shall suppose that 5,000 bees were present; they also occupied the 500 auger-holes, but only after intense competition, resulting in 90% of the population failing to contribute progeny to the next generation.

The intensity of the competition varied greatly but the density of the population in the next generation remained the same. Apparently the idea of competition does not help to explain the density of the population which was determined in each instance by the number of nesting sites.

By imagining that the bees behave differently we can also use them to illustrate the broader meaning in which competition includes the harm that one animal may do to another when both seek a common resource that is not in short supply. Suppose, now, that the bees happen to be the sort that fight viciously over nesting sites whether the nesting sites are in short supply or not. Suppose that there are 500 sites but only 200 bees. Yet they fight until many have been killed or disabled. Suppose 100 remain. They use 100 holes for nests leaving 400 holes unoccupied. Even though the nesting places are not in short supply the bees are killed while fighting for nests; so this must be called competition according to Birch's definition. Clearly, the intensity of the fighting influences the numbers in the next generation. So here is an example of 'control by competition'.

This is true but is it worth mentioning? If you tell me that 'competition' determined the number of bees you have told me nothing that is precise or important; because 'competition' is a panchreston; if you tell me how the numbers are determined by fighting I understand the situation much better. Similarly, in discussing Ullyett's experiments with blowflies or Pitelka's observations on humming birds, if you say that the blowflies or the humming birds 'competed' you have told me scarcely anything, but if you say that one sort of maggot attacked and killed the other, or that one sort of humming bird drove the other away by 'aggressive displays', I know much more precisely what you mean. In sections 6.221, 6.223, and 8.2 I have described a number of relationships between different species which come within the second meaning in Birch's definition of competition. For example, the rat which serves as a reservoir for typhus must be said to compete with man if it seeks to share the same food even though, in the presence of plenty of food for both, the competition consists merely in being a reservoir for a common pathogen. Yet with some other sort of animal, say a squirrel, which did not eat the same food nor live in the same place as a man there would be no competition no matter how effective the squirrel might be as a reservoir for typhus. The examples that are mentioned in this book represent only a small proportion of the diversity of relationships that will fit Birch's second meaning for competition. Indeed the diversity is such that there seems to be nowhere to draw a clear-cut distinction between competition and no-competition except at the point where the resource that is shared in common is in short supply. It seems that to lump all this diversity together under the panchreston 'competition' does not help us to understand the problem at all. It may even be misleading: see the comment quoted from Weiss (sec. 1.3).

Perhaps this argument may be reinforced by considering the units in which competition is measured. In the imaginary examples with the bees the intensity of the competition was measured by the *proportion* of the original population that was still surviving after the competition had occurred. This is the only way to measure it because no other unit is appropriate. That is, competition cannot be measured independently of survival-rate. Consequently there is no sense in which it is proper to say that one determines the other; the two words are merely different ways of saying the same thing. For example, we can say that the food was in short supply so the survival-rate was 10%. Or, we can say the food was in short supply so the competition was such that 10% of the animals survived. The two statements replace each other precisely; so neither adds anything to the other.

9.33 *The place of the model, the hypothesis, the null hypothesis, the experiment and the theory in scientific method*

My chief criticism of the 'theory' of density-dependent factors is that it is unscientific because its authors have depended too much on insight and deduction and not enough on experiment and observation (secs. 9.3, 9.31). I have, therefore, in this section, set out the essential features of the scientific method so that I may, by referring back to it, make my criticism as particular as possible. I shall introduce the subject by referring to a simple experiment which is described in Part II.

In section 2.1, I discussed my idea of environment. In section 11.2 I describe experiment 11.22 with the pupae of *Tribolium confusum* and *T. castaneum* in which I measure their length of life at the 'lethal low' temperature of 5°C. The *idea* and the *experiment* are linked by the following epistemological steps.

In the first place I am led to do an experiment with temperature because temperature enters into my idea of environment. In the second place I have to arrive at a particular proposition about the influence of temperature on *Tribolium* which must be such that it can be verified by experiment. I do this by a mixture of experience, insight and logic. It so happens that I do not have much experience with *Tribolium* but I do know that *Calandra*, which lives in similar places and is supposed, like *Tribolium*, to have originated in the tropics, can survive about 7 days' exposure to 5°C (sec. 5.12). I argue from this that *Tribolium* is likely to be about as cold-hardy as *Calandra*. I also happen to know that *T. castaneum* grows relatively more quickly at higher temperatures than *T. confusum* (exp. 11.11) and I argue that because *T. castaneum* is better adapted to high temperatures than *T. confusum* it will be more easily killed by cold than *T. confusum*. In this way I arrive at the following proposition which it is my intention to test empirically: both species of *Tribolium* will be killed by about a week's exposure to 5°C but *T. castaneum* will die more quickly at 5°C than *T. confusum*.

It is convenient to give a name to this stage in the scientific method because, by giving it a name, we can emphasize the fact that there are other stages to follow which are necessary for the discovery of new knowledge. The name is 'conceptual model'. Conceptual model is an apt name for this stage in the scientific method because it implies that a man has used insight, experience and logic to build a picture (or model) of some part of nature which he will subsequently test by experiment to find out whether nature really is like that. Conceptual model connotes not only such particular statements as '*Tribolium* will live for 7 days at 5°C' but also broad generalizations like 'the environment of any animal comprises four major components'. The truth of the particular statement may be tested by a single experiment. The *truth* of the broad proposition which expresses a basic idea may never be tested in this way, nor need it be; but the results of a large number of such experiments may provide a test of its *usefulness*. A familiar example from physics of how a flash of insight provided a new idea (a new way of looking at old familiar experiences) which led to much progress is the idea that 'light travels in straight lines'. The whole of geometrical optics stems from this idea and the experiments that it has led to. An earlier idea about light was the one held by the Greek philosophers that we can perceive illuminated objects by virtue of invisible 'antennae' that stretch out from our eyes. There was nothing to stop any one building a perfectly logical model on this base but it proved useless because it did not lead to useful experiments.

Between the model and the experiment comes the 'null hypothesis'. The null hypothesis is got by deducing from the model a positive conclusion (such as *T. castaneum* will die more quickly at 5°C than *T. confusum*) and reframing it as a particular sort of negative statement which asserts that there is no difference between two sets of measurements, or between one set of measurements and some hypothetical quantity with which they are to be compared. The positive statement which is the converse of a null hypothesis is often referred to simply as a hypothesis. This is all right provided that its function is not confused with that of the null hypothesis.

In the example that I am using it is convenient to state two null hypotheses (the skilful experimenter often contrives to test many null hypotheses in one complex experiment). These are: (*a*) There is no difference in the death-rate either of *T. castaneum* or of *T. confusum* whether they are living at 5°C or 27°C. (*b*) There is no difference between the length of life at 5°C of *T. castaneum* and *T. confusum.*

Then comes the *experiment*. The express and only purpose of any scientific experiment is to disprove a *null hypothesis*. Fisher (1947, p. 16) stated the purpose of the experiment clearly: 'It should be noted that the null hypothesis is never proved or established, but is possibly disproved, in the course of the experimentation. Every experiment may be said to exist only in order to give the facts a chance of disproving the null hypothesis.' If the null hypothesis is disproved at a sufficiently high level of probability we are prepared to treat its converse (that is the positive statement taken over from the model) as a 'fact' and to act upon it. There are two components to the uncertainty of the facts that are derived from experiments. One is inversely associated with the degree of probability with which the null hypothesis was disproved. The other is inversely associated with the precision with which the null hypothesis was first formulated and subsequently converted to its converse. These causes for uncertainty may never be eliminated. Consequently the facts that are established by experiment can never be stated with absolute certainty. The nearest that a scientist can ever get to absolute certainty is to disprove a precisely formulated null hypothesis at a high level of probability (Agassi, 1959). In our experiment with *Tribolium* we disproved both null hypotheses at a probability of less than 1% (exp. 11.22).

When a sufficient number of facts have been established it may be possible to make a general statement that is consistent with all the facts. This may be called a theory. For example in section 11.2, experiment

11.22, paragraph (*f*), with no more than three facts at our command, we were able to make a start by formulating the simple theory that *T. confusum* was more hardy than *T. castaneum.* A theory may never be stated with absolute certainty because it embraces the uncertainty which is inherent in the facts that it summarizes; and in addition, there is always the possibility that some other generalization will be found which is also consistent with the same set of facts. For example, the results of Waddington's (1956) experiment with *Drosophila* might, if they had stood alone, have been held to be consistent with either a theory of blending inheritance (Lamarck) or of particulate inheritance (Mendel). (In this instance the choice became clear from a consideration of certain other facts that had been established in other experiments.)

If, in the first place, our insight was keen and our logic sound, the theory that we ultimately arrive at may bear considerable resemblance to the model with which we started. The theory or some part of it may inspire a further cycle of imaginative and logical thinking which may lead to a new or revised model. (In the example that I have used in this discussion our first simple theory led to speculation about the evolution of the genus *Tribolium*; experiment 11.22.) Scientific knowledge always grows by this cyclical process. Because the process is cyclical the distinction between the stages may tend to become blurred unless one remains alert to distinguish between them.

Perhaps the fundamental distinction that has to be made is between two sorts of knowledge, depending on the method by which the knowledge is attained. The knowledge which is the conclusion of a logical argument is different from the knowledge which is the conclusion of an experiment – i.e. the converse of a disproved null hypothesis. The fruition of the first method is in abstract mathematics, philosophy and dogmatic religion. The fruition of the second method is in scientific discovery – penicillin and the hydrogen bomb.

Model-making, the imaginative and logical steps which precede the experiment, may be judged the most valuable part of scientific method because skill and insight in these matters are rare. Without them we may not know what experiment to do. But it is the experiment which provides the raw material for scientific theory. Scientific theory cannot be built directly from the conclusions of conceptual models. This is the chief mistake made by the adherents to the 'theory' of 'density-dependent factors' (sec. 9.31). That is why Andrewartha and Birch (1954) called this 'theory' a dogma.

9.34 *Reasons for rejecting the 'theory' of density-dependent factors*

If you look again at the seven passages quoted at the end of section 9.31 you will see that each one contains internal evidence of its dependence on insight, faith or logic. Elsewhere Nicholson (1954, p. 54) has stated explicitly his belief (*a*) that new scientific knowledge about animals may be discovered by the processes of deductive logic and (*b*) that certain 'facts' about animals, that have been established empirically, can be known with absolute certainty.

> The foregoing discourse begins with the statement of certain facts of common knowledge which are so well established as to be incontrovertible, [elsewhere, p. 59, in a similar context he speaks of 'axiomatic' facts]. The implications of these facts are examined in relation first to very simple situations, and then to progressively more complex ones, and eventually general conclusions are reached concerning population growth and maintenance *under natural conditions* [my emphasis]. Although much use is made throughout of further evidence derived from observations and experiments *the method of investigation is essentially deductive reasoning* [my emphasis]. The approach is thus much the same as that used in earlier articles. (Nicholson, 1933; Nicholson and Bailey, 1935.)

These statements leave no room for doubt that Nicholson's writing on this subject has the status of a conceptual model, and I think that this is also true of the last two passages that I quoted in section 9.31, as well as nearly everything else that has been written on this subject.

The only appropriate criterion by which to judge a conceptual model is its usefulness. Is it true? is a meaningless question to ask about a model because the only test of truth about nature lies in experiments – not in inspiration or logic (Fisher, 1947; Russell, 1956, p. 156). Andrewartha (1957, 1959) discussed Nicholson's model of density-dependent factors and showed that it was of little value because the hypotheses that arose from contemplation of this model were useless because they cannot be tested empirically. They cannot be tested empirically because they are stated in terms of 'equilibrium density', 'ultimate upper and lower limits' or 'reasons for non-extinction' which have no reality relative to the numbers that an ecologist can actually count in a particular place during a particular period. It is clear that this conceptual model has little chance of helping the ecologist to explain the distribution and abundance of the species that he studies.

Of course, the fact that we have to reject the extreme models that have been built on the idea of density-dependent factors does not necessarily mean that we cannot make use of Howard and Fiske's original idea that environment may be divided into density-dependent and density-independent factors. Viewed purely as a descriptive device this idea seems attractive because nearly all the 'events' that influence the density of a natural population may be classified into one or other of these categories or an extension of them (Andrewartha, 1957). In particular the reciprocal relationship which exists between the density of a population and the amount of some expendable resource that is in short supply (sec. 7.2) has often been described in terms of density-dependent factors (Huffaker, 1957). But the idea of density-dependent factors cannot be fitted into the model of environment that I described in Chapter II. This model of environment has been proved to be generally useful and there seems little merit in discarding it in favour of a narrower model which can at best explain only a small part of the variability that may be observed in natural populations. For example the relationships that are illustrated in Figures 7.02, 7.03, 9.02, and certain aspects of 9.03, 9.07 and 9.08 might be explained in terms of density-dependent factors, but those that are illustrated in Figures 9.01, 9.04, 9.05, 9.09 and most aspects of 9.03, 9.07 and 9.08 are outside the scope of this idea.

9.4 FURTHER READING

Because I have emphasized the unique importance of the empirical part of the scientific method this does not mean that I belittle the importance of imagination, insight and logic. In science, as in mathematics, most of the great advances have been made possible by someone finding a new way of looking at old familiar facts, by finding a new way of talking about the same thing – in a word, by conceiving new ideas. Perhaps the history of ideas may be traced more easily for mathematics than for science. In any case, this has been done most delightfully for mathematics by Dantzig (1942) in his book *Number the Language of Science*. Bronowski (1951) deals more explicitly with the history of some of the basic ideas of science.

The second part of this chapter deals with a controversial subject. So it is difficult to recommend any further reading without recommending too much. The student who wishes to follow it up and has the leisure will find that the references given in the text of this chapter will lead him quite deeply into the subject.

PART II

Practical Course

CHAPTER X

Methods for Estimating Patterns of Distribution Density and Dispersal in Populations of Animals

10.1 PATTERNS OF DISTRIBUTION

This subject was discussed in sections 3.3 to 3.33.

Experiment 10.11

(*a*) Purpose

To compare the distribution of weevils *Calandra granaria* in a box of wheat with the theoretical distribution given by the Poisson series. This is a laboratory model. The null hypothesis is that there is no difference between the observed distribution of weevils and a Poisson series.

(*b*) Material

A laboratory culture of *Calandra granaria* from which 1,000 to 2,000 adult weevils may be collected (Appendix I).

(*c*) Apparatus

(i) A flat box of wood, or of some other convenient material, measuring about 100 cm. square by about 10 cm. deep.

(ii) Sufficient clean wheat to half-fill the box.

(iii) Sampling cylinders made by cutting the tops and bottoms out of cans about 7 cm. in diameter and 6 cm. deep.

(iv) Press-lid cans or screw-top jars of about the same size as the sampling cylinders or a little larger. They are used to hold the samples while they await counting.

(v) Scoops for dipping the wheat out of the sampling cylinders. They can be made from small spoons bent at right angles at the throat.

(vi) Shallow rectangular trays for tipping the samples into for counting. Aluminium cake-dishes about 20 cm. square by 3 cm. deep are satisfactory.

(*d*) Method

(i) *Establishing the colony*

Fasten a flange of foil or thin cardboard around the edges of the box to contain the weevils. Mark out a scale of about 50 divisions along each edge of the box. Tip wheat into the box until it forms a smooth layer about 5 cm. deep over the whole box. Release the weevils near the centre of the box. Bury them below the surface and do not crowd or jostle them more than you have to; otherwise they may emerge and crawl over the surface instead of dispersing quietly through the wheat. Leave the weevils for about a week to distribute themselves through the wheat before proceeding with the experiment.

(ii) *Sampling*

Stretch thread or fine string across the top of the box from opposite edges to mark out a rectangular grid of about 100 squares. Take a sample from the centre of each square by pushing a sampling cylinder firmly against the bottom of the box and scooping out the wheat into a container. The last few weevils may be picked up with forceps or brush and dropped into the container.

(iii) *Counting*

Tip the wheat, a little at a time, from the container to the counting tray. Shake it to make sure that all the weevils are seen. Count them and then return the wheat and the weevils to about the same place in the box that they were taken from.

(*e*) Results

In our experiment we used 72 quadrats instead of 100. Table 10.11*a* sets out the numbers of weevils that were found in the 72 quadrats. The mean number per quadrat is,

$$\bar{x} = \frac{Sx}{n} = \frac{177}{72} = 2{\cdot}458$$

The variance is calculated from the expression,

$$s^2 = \frac{S(x-\bar{x})^2}{n-1} = \frac{1}{n-1}(Sx^2 - Sx.\bar{x})$$

$$= \frac{909 - 435{\cdot}125}{71} = 6{\cdot}674$$

TABLE 10.11*a*

The numbers of weevils in the samples, arranged as they occurred in the box

12	2	0	7	6	1	2	9
2	5	1	4	1	1	0	3
6	2	0	2	2	0	2	2
8	2	0	0	1	1	0	2
4	2	1	1	4	1	3	2
3	1	1	1	3	1	1	1
0	1	1	2	2	1	0	2
3	8	2	3	1	1	1	1
2	2	3	2	12	2	2	7

The frequencies in the Poisson series which has $N = 72$ and $\bar{x} = 2{\cdot}458$ is given by simplifying,

$$\frac{72}{2{\cdot}718^{2{\cdot}458}}\left(1,\ 2{\cdot}458,\ \frac{2{\cdot}458^2}{2!},\ \frac{2{\cdot}458^3}{3!},\ldots\right)$$

which is 6·164, 15·151, 18·621, . . . (see Table 10.11*b*, col. 5). In column 6 of Table 10.11*b* $(O - E)$ is the difference between columns 2 and 5. The sum of the quantities in this column is χ^2. Because the expected frequencies were calculated from the total and the mean, χ^2 has $n - 2 = 4$ degrees of freedom.

(*f*) Conclusions and comment

The value $\chi_4{}^2 = 14{\cdot}41$ provides odds of rather better than 100 to 1 against the null hypothesis. We conclude that the weevils were distributed non-randomly over the box of wheat. The large value for the variance relative to the mean indicates that the departure from randomness was in the direction of excessive aggregation. And an inspection of Table 10.11*a* indicates that the weevils had a strong tendency to congregate around the edges of the box and in the corners. Considering that there were no more than 2,000 weevils in a box 100 cm. square it seems

TABLE 10.11*b*

The observed frequencies compared with the theoretical frequencies calculated from the Poisson series

Number of weevils per sample (x)	*Number of samples* (f)	fx	fx^2	*Expected frequencies (Poisson)*	$\frac{(O-E)^2}{E}$
0	9	0	0	6·164	1·31
1	22	22	22	15·151	3·10
2	21	42	84	18·621	0·34
3	7	21	63	15·255	4·47
4	3	12	48	9·375	4·34
5	1	5	25		
6	2	12	72		
7	2	14	98	7·420	0·90
8	2	16	128		
9	1	9	81		
12	2	24	288		
Total	72	177	909	71·99	14·41

Ratio $\frac{s^2}{\bar{x}} = \frac{6 \cdot 674}{2 \cdot 458} = 2 \cdot 72; \chi_4^2 = 14 \cdot 41$

likely that this was a response to some physical stimulus rather than to crowding but we did not follow this observation up.

Experiment 10.12

(*a*) Purpose

To compare the distribution of the scale insect *Saissetia oleae* on an olive bush with the theoretical distribution calculated from the Poisson series. The null hypothesis is the same as in experiment 10.11.

(*b*) Material

Any large shrub or small tree carrying a not too dense population of scale insects. For our exercise we chose a stunted olive bush about 2 metres high.

(*c*) Apparatus

Secateurs and a basket in which to collect and store the twigs until the scale insects have been counted.

(*d*) Method

Choose some suitable length of twig such that the average twig is likely to have not more than two or three scale insects on it. Clip a convenient number (between 100 and 200) twigs from the bush working systematically over the bush. We took the terminal 10 cm. of every branch up to a height of about 1½ metres. Count the scale insects on each twig and arrange them in a frequency table as in Table 10.12*a*.

(*e*) Results

The results have been analysed in the same way as those in experiment 10.11; they have been presented in Table 10.12*a* which has the same form as Table 10.11*b*.

TABLE 10.12*a*

The observed frequencies compared with the theoretical frequencies calculated from the Poisson series

Number of scale insects per twig (*x*)	*Number of twigs* (*f*)	*fx*	*fx*²	*Expected frequencies* (*Poisson*)	$\frac{(O-E)^2}{E}$
0	187	0	0	82·69	131·5
1	63	63	63	109·54	19·8
2	18	36	72	72·55	41·0
3	16	48	144	32·04	8·0
4	9	36	144	10·61	
5	2	10	50	2·81	
6	2	12	72	0·62	
7	5	35	245	0·12	
8	3	24	192	0·02	
13	1	13	169	0·00	12·8
17	1	17	289	0·00	
19	1	19	361	0·00	
25	1	25	625	0·00	
34	1	34	1156	0·00	
40	1	40	1600	0·00	
Total	311	412	5182	311·00	213·1

$\bar{x} = 1{\cdot}3247$; $s^2 = 14{\cdot}96$; ratio $\frac{s^2}{\bar{x}} = 11{\cdot}3$; $\chi_3^2 = 213$

(*f*) Conclusions and comment

The discrepancies between observed and expected frequencies are so large and obvious that there was scarcely any need to carry out the calculation of χ^2 – the result was a foregone conclusion. The value of over 200 for χ_3^2 is well out of the range of Table IV in Fisher and Yates indicating odds of thousands to one against the null hypothesis. We conclude that the scale insects were distributed non-randomly: there are far too many twigs with no scale insect on them, and individual twigs with numbers as high as 40. This is a good example of the way in which populations of insects tend to be distributed as 'local populations' (sec. 9.0).

The results set out in Tables 3.04 and 3.05 were also got by students as a practical exercise. Marine forms like *Parcanassa* make excellent material for a class exercise. But I have quoted the exercise that we did with *Saissetia* because it shows how this method may be extended to less conventional situations than a flat box of wheat or a flat stretch of mud.

Experiment 10.13

(*a*) Purpose

To compare the pattern of distribution in two colonies of *Paramoecium*. We have two null hypotheses. (i) That the distribution of the *Paramoecium* in our samples is not significantly different from the Poisson series. (ii) That there is no difference in the distributions of the *Paramoecium* in the two dishes, i.e. the correlation coefficient between them does not differ significantly from zero.

(*b*) Material

A culture of *Paramoecium* (Appendix I).

(*c*) Apparatus

(i) Two flat glass dishes. We use petri dishes 22 cm. in diameter.

(ii) Two sheets of paper to go under the petri dishes with a square grid ruled on them. The scale of the grid should be such that the bottom of the dish covers about 100 squares.

(iii) A sampling rod made by rounding one end of a 5-mm. glass rod in a flame.

(iv) Microscope slides and a binocular dissecting microscope for counting. *Paramoecium* may be counted under a magnification of 12·5 times.

(*d*) Method

(i) Tip medium containing *Paramoecium* into the two petri dishes until it is about half an inch deep. Take a few preliminary samples by dipping the rod into the medium and touching the drop of medium on to a microscope slide. If necessary adjust the density of the population or the sampling rod so that not more than 6 *Paramoecium* are likely to be taken, on the average, in one sample. Place the dishes on a bench in front of a window in bright light but not in direct sunlight. Place the paper grids under the dishes so that the grids are orientated parallel to each other. Leave the dishes undisturbed for several hours to give the *Paramoecium* time to distribute themselves naturally.

(ii) Take a sample from above each square in the grid by dipping the rod into the medium and touching the drop on to a slide. Keep a record of the squares from which the samples were taken so that a table like Table 10.13*a* may be constructed to compare the numbers in corresponding squares in the two dishes.

(*e*) Results

(i) The results are given in Table 10.13*a*. The upper numbers in each cell of the table refer to dish A and the lower numbers to dish B. The results from the two dishes have first to be tested against the Poisson series to see whether the *Paramoecium* are distributed at random. If either distribution conforms to the Poisson series there is no point in carrying the analysis any further. But if both distributions are non-random and 'patchy' then it may be appropriate to ask whether the pattern of patchiness in one dish resembles that in the other as it would if both were related to the same external cause such as the position of the bright window or the edge of the dish, etc. On the other hand if the two patterns are not similar we might infer that they have arisen in response to circumstances within the dish – the distribution of the food may be patchy, or perhaps *Paramoecium* is gregarious, forming clusters independent of any external influence.

(ii) The frequencies shown in Table 10.13*b* may be constructed from the raw data given in Table 10.13*a*. From the statistics given at the foot of the table it is clear that the distributions are non-random and patchy.

Because n is large P cannot be found directly from the tables of χ^2. Instead we use the table of t having first calculated,

$$t_\infty = \sqrt{2\chi^2} - \sqrt{2n - 1}$$

But the distribution of χ^2 corresponds to only one tail of the distribution of t. So it is necessary to halve the values of P given in the table of t (Fisher and Yates, 1948, Table III).

(iii) The degree of association between the two distributions may be measured by calculating the correlation coefficient:

$$r = \frac{S(x - \bar{x})(y - \bar{y})}{\sqrt{S(x - \bar{x})^2 \times S(y - \bar{y})^2}}$$

TABLE 10.13*a*

Numbers of *Paramoecium* in samples from corresponding positions in two similar dishes

				4 6	1 10	11 11	5 8	1 4	5 8				
			4 2	2 4	5 5	11 10	7 7	3 2	7 3	9 3			
		1 1	5 5	6 6	10 10	8 8	7 5	11 13	5 10	2 9	5 11		
	4 6	5 7	5 3	5 7	4 4	9 3	7 8	8 9	8 7	4 8	7 7	7 6	
6 3	7 0	7 1	5 6	8 11	6 6	12 19	5 13	9 9	7 4	11 10	9 5	14 1	8 6
4 6	10 4	3 2	11 1	8 7	6 5	7 8	8 16	4 10	4 8	3 0	7 3	5 5	4 8
1 5	12 3	7 2	0 2	10 3	8 3	11 12	13 4	2 6	8 10	2 1	2 10	7 2	8 3
2 6	4 2	6 4	5 3	3 15	5 4	5 7	7 8	5 3	3 5	3 4	4 5	3 3	6 0
10 5	6 3	8 4	5 10	5 10	6 11	5 8	6 8	5 12	9 8	9 6	7 1	6 1	1 6
5 9	10 4	5 3	8 6	11 13	9 4	13 5	4 10	1 5	7 2	6 4	8 7	8 0	5 5
	5 12	6 5	1 2	3 10	6 5	2 3	2 5	7 2	2 4	5 5	5 1	3 7	
		4 16	1 5	2 11	6 7	6 0	3 7	5 5	7 5	4 6	2 8		
			4 4	3 6	1 4	3 3	1 4	2 2	7 2	5 2			
				6 3	0 3	4 1	8 7	0 0	0 5				

TABLE 10.13*b*

No. of Paramoecium *per sample* (x)	*No. of samples* (f)	
	Dish A	*Dish B*
0	4	6
1	10	9
2	12	13
3	12	19
4	16	17
5	28	21
6	16	15
7	19	12
8	15	13
9	7	4
10	5	12
11	7	5
12	2	3
13	2	3
14	1	0
15	0	1
16	0	2
17	0	0
18	0	0
19	0	1
Total ($Sf = n$)	156	156
$S(x - \bar{x})^2$	1190·22	2224·69
$s^2 = \frac{S(x - \bar{x})^2}{n - 1}$	7·68	14·35
Mean $\bar{x}$	5·62	5·74
$\chi^2 = \frac{S(x - \bar{x})^2}{\bar{x}}$	396	208
Ratio $\frac{s^2}{\bar{x}}$	2·6	1·3
P	$<0·001$	$<0·005$

x and y refer to the individual counts from dish A and dish B respectively. If the calculation is to be done on a machine it is convenient to use the identities:

$$S(x-\bar{x})(y-\bar{y}) \equiv Sxy - \frac{Sx.Sy}{n}$$

$$S(x-\bar{x})^2 \equiv Sx^2 - \frac{(Sx)^2}{n}$$

$$S(y-\bar{y})^2 \equiv Sy^2 - \frac{(Sy)^2}{n}$$

In the present instance the numerator,

$$S(x-\bar{x})(y-\bar{y}) = \text{the sum of } 4\times 6 + 1\times 10 + 11\times 11 + \ldots + 8\times 7 + 0\times 0 + 0\times 1$$

$$- \frac{(\text{sum of } 4+1+11+\ldots+8)(\text{sum of } 6+10+11+\ldots+5)}{156}$$

Thus

$$r = \frac{5{,}276 - 5{,}031{\cdot}5}{\sqrt{2{,}224{\cdot}7 \times 1{,}190{\cdot}2}} = 0{\cdot}15$$

There are 154 (i.e. $n - 2$) degrees of freedom for r which is beyond the range of Table VI in Fisher and Yates. So we use the table of t instead, having calculated t from the equation:

$$t_\infty = \frac{r\sqrt{n-2}}{\sqrt{1-r^2}} = \frac{0{\cdot}15 \times \sqrt{154}}{\sqrt{0{\cdot}85}} = 2{\cdot}018$$

In the bottom line of Fisher and Yates' Table III, P is given as between 0·05 and 0·02 for this value of t.

(*f*) Conclusion and comment

In both dishes A and B the *Paramoecium* were distributed more patchily than might have been reasonably expected from a random scatter. The discrepancies were highly significant in each instance. Consequently it was pertinent to compare the distributions in the two dishes to see whether any external influence (i.e. external to the dish) could be detected that was common to both dishes. As a first step the correlation coefficient was calculated. This turned out to be small, but significant at the 5% level of probability. That is, there was only one chance in 20 that this degree of association would have occurred by chance. Nevertheless the degree of association was slight leading to the conclusion that whatever external stimulus may have been operating its influence was small relative to the internal influences.

The correlation coefficient is the ratio of the covariance between x and y to the geometric mean of their variances. Consequently any modification of technique which reduced the variances might enhance one's chance of demonstrating a significant correlation. For example one line of approach would be to make the media in the dishes more homogeneous.

10.2 METHODS FOR ESTIMATING THE DENSITY OF A POPULATION

The theoretical aspects of this subject were discussed in sections 3.1 to 3.23.

Experiment 10.21

(*a*) Purpose

To estimate the number of weevils *Calandra granaria* living in a box of wheat. The wheat forms a layer 6 cm. deep in a box that is about 100 cm. square and 10 cm. deep. (It is the same population that was used in experiment 10.11.) This experiment is a laboratory model of the quadrat method of estimating density (sec. 3.22).

(*b*) Material

The same population of weevils that was used in experiment 10.11.

(*c*) Apparatus

The same as for experiment 10.11 and in addition a table of random numbers, e.g. Fisher and Yates (1948, Table XXXIII).

(*d*) Method

Take a convenient number, say 50, samples from the box. It is essential for this method that the samples be taken at random from the whole area of the box, i.e. any sample must be allowed an equal chance of coming from any part of the box that has not already been sampled. The best way to distribute the samples at random is to use the scale marked on the edges of the box as the axes of a graph and to locate each sample by choosing a pair of co-ordinates from the table of random numbers. The sample is taken as in experiment 10.11 by pressing the sampling cylinder against the floor of the box and scooping the wheat into a suitable container to await counting.

(*e*) Results

In 50 samples we got 118 weevils.

The mean number per sample: $\bar{x} = \frac{118}{50} = 2.36$

The population variance: $s^2 = \frac{S(x - \bar{x})^2}{n - 1} = 6.94$

The variance of the mean: $\frac{s^2}{n} = \frac{6.94}{50} = 0.1388$

The standard deviation of the mean: $\frac{s}{\sqrt{n}} = 0.373$

The fiducial limits of the mean are calculated from the equation:

$$\text{Fiducial limits} = \bar{x} \pm t \times \frac{s}{\sqrt{n}}$$

From Table III of Fisher and Yates for $P = 0.05$ and $n = 49$, $t = 1.68$.

So $$\bar{x} \pm t \times \frac{s}{\sqrt{n}} = 2.36 \pm 1.68 \times 0.373$$

$$= 1.73,\ 2.99$$

(*f*) Conclusion and comment

The surface area of the box was 338·6 times the area of the sampling cylinder. So there is one chance in 20 that the population was smaller than 1.73×338.6 or larger than 2.99×338.6. In other words, the odds are 19 : 1 that there were between 586 and 1,012 weevils in the box at the time of our estimate.

With 50 quadrats, equivalent to about one-seventh of the total area, one might have expected a more precise estimate of the population than this. The lack of precision was probably associated with the patchiness of the distribution (Table 10.11*b*).

Experiment 10.22

(*a*) Purpose

To compare the relative densities of two populations of *Paramoecium* in two large petri dishes. This experiment may be regarded as a laboratory model of the field method of 'simple trapping' (sec. 3.1). The null hypothesis is that there is no difference in the densities of the two populations.

(*b*) Material

A culture of *Paramoecium* (Appendix I).

(*c*) Apparatus

(i) Two large petri dishes about 22 cm. in diameter.

(ii) Two sheets of graph paper at least as large as the petri dishes.

(iii) Culture medium suitable for *Paramoecium* (Appendix I).

(iv) Glass sampling rods made by heating one end of a short length of 5-mm. glass rod in a flame until the end is slightly bulbous.

(v) Microscope slides and a binocular dissecting microscope for counting *Paramoecium*.

(vi) A table of random numbers.

(*d*) Method

Place a sheet of graph paper under each petri dish. Half-fill the dishes with medium and transfer an unknown number (preferably several thousand) *Paramoecium* to the dishes.

Take 50 (or thereabout) random samples from each dish. Take the samples by dipping the sampling rod into the medium and then touching the drop of medium on to a microscope slide. Randomize the position of the samples by using the table of random numbers and the graph paper as in experiment 10.21. Use a different set of random co-ordinates for each dish. Count the *Paramoecium* in each sample.

(*e*) Results

The mean number of *Paramoecium* per sample, the standard deviation of the mean and the fiducial limits for the estimate of the mean per sample may be calculated as in experiment 10.21. The final step of estimating the actual number of *Paramoecium* in the dish cannot be carried out because we do not know what proportion of the whole was represented in one sample.

The negative (or 'null') hypothesis that the density of the populations in the two dishes did not differ from one another may be tested by the t test. If we write $\bar{x}_1$ for the mean number of *Paramoecium* per sample in dish A and n_1 for the number of samples, and $\bar{x}_2$ and n_2 for the same statistics for dish B, then

$$t = \frac{\bar{x}_1 - \bar{x}_2}{\text{standard deviation of the difference between means}}$$

The standard deviation of the difference between means may be calculated from,

$$\text{S.D. diff.} = \sqrt{\frac{S(x_1 - \bar{x}_1)^2 + S(x_2 - \bar{x}_2)^2}{n_1 + n_2 - 2}} \times \sqrt{\frac{n_1 + n_2}{n_1 \times n_2}}$$

In one experiment that we did

$$\text{S.D. diff.} = \sqrt{\frac{260{\cdot}76 + 288{\cdot}00}{154 + 160 - 2}} \times \sqrt{\frac{154 + 160}{154 \times 160}}$$

$$= 0{\cdot}473$$

$$t = \frac{1{\cdot}34 - 0{\cdot}75}{0{\cdot}473} = 1{\cdot}25$$

t has $n_1 + n_2 - 2$ degrees of freedom. From Table III of Fisher and Yates (1948) it may be seen that for $t = 1{\cdot}282$ and $n = 120$, $P = 0{\cdot}2$.

(*f*) Conclusions and Comment

The odds are rather less than 4 to 1 against the null hypothesis. Usually we require the odds against a null hypothesis to be at least 19 to 1 before we accept the difference as 'significant'. So we have failed to disprove our null hypothesis. The conventional way of describing the results of the present experiment would be to say that the difference was non-significant.

This experiment measured the relative but not the absolute densities of the two populations because we did not know what proportion of the whole medium was contained in the drop that was lifted out with the sampling rod.

Experiment 10.23

(*a*) Purpose

To estimate the number of weevils *Calandra granaria* in a flat box of wheat by the method of capture, marking, release and recapture (Jackson's positive method – sec. 3.23). This is a laboratory model: the method of 'capturing' that is used is analogous to the method of 'direct search' that is often used in the field.

(*b*) Material

A culture of weevils from which a convenient number (say about 2,000) may be collected.

(*c*) Apparatus

The same as for experiment 10.21 and, in addition, some quick-drying cellulose lacquer for marking the weevils.

(*d*) Method

Transfer about 2,000 weevils to the flat box of wheat. Handle them gently to avoid agitating them. If they are crowded too much or handled roughly they become very active and are likely to crawl over the surface of the wheat and escape from the box instead of burrowing in and settling down. Allow the weevils several days or perhaps a week to distribute themselves naturally through the wheat. Then take samples from the wheat – 50–100 is a convenient number. Randomize the positions of the samples in the same way as in experiment 10.21. Mark the weevils with a small spot of paint and replace them in the place that they came from. For the next three to six weeks take weekly samples at random from the box and record the number of weevils captured including the number of marked ones that are recaptured. On each occasion return the weevils to about the same place that they came from.

(*e*) Results

In the experiment that I report below we had time for only 3 recaptures. The results are given in Table 10.23*a*.

TABLE 10.23*a*

The number of weevils marked and recaptured and the number captured on each occasion

Date	*March 7*	*March 14*	*March 21*	*March 28*
No. marked	498			
Total captured		110	380	184
Including number recaptured		20	48	17

The values of y, r, and a were calculated from the formulae given in section 3.23.

They were $y_1 = 3{\cdot}651$; $y_2 = 2{\cdot}536$; $y_3 = 1{\cdot}855$; $r_+ = 0{\cdot}7097$; $a = 5{\cdot}067$.

The estimate for the number of weevils in the box on March 7 was then calculated as,

$$N = \frac{10{,}000}{a} = 1{,}974$$

(*f*) Conclusion and Comment

Although we recaptured on only three occasions we were able, at the beginning of the experiment, to mark about 20% of the population. With such a high proportion marked, and with no evidence from the individual samples that the marked had not distributed themselves similarly to the unmarked, it is likely that this is a reasonably good estimate. There are statistical methods for measuring the homogeneity of the distribution of marked and unmarked, and also for calculating the variance of the estimated population, which would allow us to state precisely how good our estimate is. But these methods are beyond the scope of this book (sec. 3.4).

Experiment 10.24

(*a*) Purpose

To estimate the number of snails *Helix* sp. in a garden by the method of capture, marking, release and recapture (Jackson's negative method).

(*b*) Material

A population of snails living naturally in a garden. The garden was mostly shrubbery. It measured about 150 feet by 50 feet.

(*c*) Apparatus

All that is required is a suitable material for marking the snails. We tried a variety of cellulose and other lacquers which would not stay on the snails during wet weather but we found a proprietary line of nail-polish that was durable. We found six shades that we could distinguish on the recaptured snails.

(*d*) Method

At certain seasons of the year (at Adelaide usually during September, October) the snails may be trapped quite readily under boards or inside inverted flower-pots especially if they are baited with a little metaldehyde. But at other times of the year very few are caught in the traps. Even during the spring their behaviour is erratic and it is better, though more laborious, merely to search for them. Each snail, as it is found, is marked with the colour appropriate to that date and released. The intervals between the days on which the snails are marked and released must be constant. About one week is a convenient interval because it allows the marked snails time to distribute themselves similarly to the unmarked ones, and it does not prolong the experiment unduly. It is desirable to mark and release on at least six occasions; the minimum number that is practicable is three. Some snails may be caught several times. They should be marked with the appropriate mark on each occasion.

(*e*) Results

The results are given in Table 10.24*a*. The value for r_-, as calculated from equation 3.3, was 0·8889; the values for y_{-1} to y_{-7} and a were calculated from equations 3.2 and 3.5 respectively. The estimated population on June 19 is given by:

TABLE 10.24*a*

Date	t_{-7} *May 1*	t_{-6} *May 8*	t_{-5} *May 15*	t_{-4} *May 22*	t_{-3} *May 29*	t_{-2} *June 5*	t_{-1} *June 12*	t_0 *June 19*
No. captured and marked	31	57	73	108	97	104	97	
No. recaptured on June 19	4	7	10	15	16	20	23	
Total captured on June 19 (including recaptures)								1,219
y	0·950	1·079	1·119	1·167	1·334	1·607	1·923	
a								2·1075

$$P_0 = \frac{10{,}000}{a} = \frac{10{,}000}{2{\cdot}1075} = 4{,}745$$

I mentioned in relation to the preceding experiment that it is outside the scope of this book to calculate the variance of this estimate. The fact that we recaptured on 7 occasions and were able on t_0 to see 95 of the 567 snails (marks) that had been marked on previous occasions suggests that this is likely to be a reasonably good estimate.

Experiment 10.25

(*a*) Purpose

To estimate the trends in the density of a population of *Tribolium castaneum* living in a flat box of flour. It is inended to use the method of capture, marking, release and recapture; Jackson's positive and negative methods will be combined to construct a trellis diagram (sec. 3.23).

(*b*) Material

A colony of *T. castaneum* containing all stages from eggs to adults. A colony that has been cultured as described in Appendix I, and has been living at 27°C in 200 gm. of flour-medium for 8 weeks is suitable.

(*c*) Apparatus

(i) A flat box about 75 cm. square and 10 cm. deep.

(ii) Enough flour-medium to make a layer about 5 cm. deep in the box.

(iii) A number of glass specimen tubes about 70 mm. by 15 mm. – for taking samples.

(iv) Some white paint or lacquer to mark the beetles.

(v) A binocular dissecting microscope.

(vi) A cylinder of carbon-dioxide and suitable apparatus for holding the beetles under light anaesthesia while they are marked under the microscope. A buchner funnel mounted in place of the stage of the microscope with CO_2 led into it from below, works quite well.

(vii) A table of random numbers.

(*d*) Method

(i) About a week before the experiment is due to start tip the culture of *Tribolium* into the box and half-fill the box with flour-medium.

(ii) Samples may be taken by pushing a glass specimen tube upside down into the medium and lifting it out. The beetles are sieved out from the medium, counted, marked and scattered in the vicinity of the same place as they came from.

(iii) The position of the samples may be randomized by drawing, from a table of random numbers, pairs of numbers to serve as co-ordinates which locate the sample with respect to scales that have been marked along the edges of the box.

(iv) The number of samples depends on the number of beetles taken relative to the total population. On the first occasion it may suffice to take enough samples to yield between 100 and 200 beetles. On subsequent occasions it is best to continue sampling until about 5 to 10% of the most recently marked ones have been recaptured.

(v) The beetles may be marked, while they are under the anaesthetic, with the extreme tip of a fine sable-hair brush (artists No. 00). There are 9 positions that may be used to indicate the date: left and right side of the thorax, anterior, middle and posterior thirds of each elytron, and if a ninth position is needed a spot of a different colour may be put centrally on the thorax.

(vi) With 9 positions available the marking and releasing may be done on 9 occasions. The occasions must be evenly spaced; a week is a convenient interval. Each beetle that is captured must be marked in the position appropriate for that date, irrespective of how many other marks it may be carrying. In counting recaptures it is the marks that are counted.

(*e*) Results

The results of one experiment are set out in Table 10.25*a*. The table is a trellis diagram like those discussed in section 3.23 except that it has been re-orientated. The figures in the body of the table are the number of beetles recaptured. Each horizontal row refers to the number marked on one date, as indicated in the left-hand margin, and recaptured on subsequent dates, as indicated along the top. Each vertical column relates to the numbers recaptured on one day, as indicated along the bottom, of those that were marked on preceding days, as indicated in the left-hand margin. Thus each row provides the raw material for estimating the size of the population (on the day of release) by Jackson's positive method; and each column provides the raw data for estimating the size of the population (on the day of recapture) by Jackson's negative method.

The raw data for recaptures must be corrected to allow for the varying numbers captured and marked on different occasions. This is done by calculating,

$$y_n = R_n \times \frac{100}{C_m} \times \frac{100}{C_n}$$

where R_n is the number recaptured on date n

C_m is the number marked on the date of release

C_n is the total captured (including recaptures) on date n.

The corrected values are shown in Table 10.25*b*. From each row in which there are 3 or more entries a value for r_+ is calculated and from each column with 3 or more entries a value for r_-. Equation 3.4 may be used if there are more than 4 entries, otherwise equation 3.3. Equation 3.5 is then used to calculate a for each row and column, and N is got by dividing 10,000 by a.

In Table 10.25*b* the recaptures that are arranged along any diagonal

TABLE 10.25*a*

Crude recaptures. The numbers in each row indicate the number of beetles recaptured on successive dates of those that were marked on the date indicated in the left margin. The columns indicate the numbers recaptured of those that were marked on the dates shown in the left margin (negative method)

Date and number marked	*Date recaptured*								
	Oct. 29	*Nov. 3*	*Nov. 8*	*Nov. 13*	*Nov. 18*	*Nov. 23*	*Nov. 28*	*Dec. 3*	*Dec. 8*
Oct. 29	117	6	9	7	9	3	3	6	5
Nov. 3		102	11	3	7	3	5	2	8
Nov. 8			213	10	11	12	6	8	7
Nov. 13				118	6	3	7	0	3
Nov. 18					145	11	6	4	4
Nov. 23						130	9	6	7
Nov. 28							147	6	6
Dec. 3								134	13
Dec. 8									
Number captured	117	102	215	119	145	131	147	137	179
Date captured	Oct. 29	Nov. 3	Nov. 8	Nov. 13	Nov. 18	Nov. 23	Nov. 28	Dec. 3	Dec. 8

The figures in the body of the table are corrected recaptures (y) which have been calculated from the expression,

$$y_n = R_n \times \frac{100}{C_m} \times \frac{100}{C_n}$$

The values for R_n, C_m and C_n are given in Table 10.25*a*'

Date marked and released	*Date recaptured*									*N* (*positive method*)
	Oct. 29	*Nov. 3*	*Nov. 8*	*Nov. 13*	*Nov. 18*	*Nov. 23*	*Nov. 28*	*Dec. 3*	*Dec. 8*	
Oct. 29		5·0	3·6	5·0	5·3	2·0	1·7	3·7	2·4	1,890
Nov. 3			5·0	2·5	4·7	2·2	3·3	1·4	4·4	(4,333)
Nov. 8				3·9	3·6	4·3	1·9	2·7	1·8	2,059
Nov. 13					3·5	1·9	4·0	0	1·4	3,332
Nov. 18						5·8	2·8	2·0	1·5	3,322
Nov. 23							4·7	3·4	3·0	1,801
Nov. 28								3·0	2·3	
Dec. 3									5·4	
Dec. 8										
N (*neg. method*)				(6,369)	3,164	2,053	1,955	(4,550)	1,998	
Date	Oct. 23	Nov. 3	Nov. 8	Nov. 13	Nov. 18	Nov. 23	Nov. 28	Dec. 3	Dec. 8	

were all taken at the same interval after release. This consideration may lead to the following argument. Let us imagine that we might have recaptured the beetles marked and released on, say November 8, immediately after they had been released. Let us call the corrected value of this imaginary recapture y_0; also let us call the corrected value for the actual recapture on November 13 (of beetles marked on Nov. 8) y_1. Similarly let the recaptures on November 8 and 13, of beetles marked on November 3, be called x_1 and x_2 respectively. Now y_1/y_0 is an estimate of r_+ (sec. 3.23) for the period November 8 to 13; similarly x_2/x_1 is an estimate of r_+ *over the same period*. It is reasonable to equate these two independent estimates of r_+. Thus:

$$\frac{y_1}{y_0} = \frac{x_2}{x_1}$$

$$\therefore y_0 = \frac{x_1 y_1}{x_2}$$

This is generally true in the circumstances in which . . . $x\ y\ z$ denote consecutive dates of marking, and $x_1\ x_2\ x_3 \ldots, y_1\ y_2\ y_3 \ldots, z_1\ z_2\ z_3$ denote consecutive dates of recapture. Therefore we have as a general estimate of y_0:

$$a = x_1 \frac{y_1 + y_2 + y_3 \ldots y_n}{x_2 + x_3 + x_4 \ldots x_n} \quad . \quad . \quad . \quad (10.1)$$

This method of estimating a seems to be theoretically sounder than the ones that we have been using because it does not depend so much on extrapolation; but it has the same defect as the others in that it depends on the assumption that the series x and y decline in geometric progressions. This is not likely to be true except in a population in which the likelihood of death is independent of age or else the age-structure of the population is invariable. Jackson (1948) described a method which seems to be more realistic because it does not depend on any of these assumptions (sec. 3.4).

The estimates of N (the number of adult *Tribolium* in the box) that were made from Jackson's positive and negative methods and from equation 10.1 are brought together in Table 10.25*c*.

(*f*) Conclusion and Comment

Table 10.25*c* gives estimates of N for each sampling day from October 29 to December 8. In the absence of any knowledge of the variance of these estimates we cannot tell how precise they are. The occurrence of

TABLE 10.25c

Details of the calculations for completing the estimates of the size of the population

Date	Positive method			Negative method			Equation 10.1	
	r_+	*a*	*N*	r_-	*a*	*N*	*a*	*N*
Oct. 29	0·943	5·290	1,890					
Nov. 3	0·951	2·308	(4,333)				4·958	2,017
8	0·884	4·858	2,059				4·919	2,033
13	0·758	3·001	3,332	1·17	1·570	(6,369)	2·946	3,394
18	0·913	3·010	3,322	1·15	3·161	3,164	5·801	(1,723)
23	0·790	5·552	1,801	0·842	4·872	2,053	10·219	(979)
28				0·902	5·112	1,955	3·892	2,569
Dec. 3				0·940	2·198	(4,550)		
Dec. 8				0·970	5·006	1,998		

one or two aberrant values (indicated by parentheses) in each of the three series of estimates suggests that sampling errors may have been large.

The causes of the fluctuations in *N* may be inferred from inspection of Table 10.25*b*. The values of *y* given in the table really stand for percentages of marked animals recaptured in a standard sample of 100 captures. Consider any row, say the one opposite November 3. If the death-rate between November 3 and December 8 was the same for marked as for unmarked then the proportion of recaptures should remain the same along the full length of the row unless the population has been increased by the addition of unmarked ones by birth or immigration. Consequently any downward trend in the row from left to right may be taken as a measure of the increments due to births and immigration. In this experiment immigration was arbitrarily prevented and births mean the emergence of adults from pupae. Now consider any column, say the one above the recapture date of December 3. If the population has been growing by the addition of unmarked ones through births or immigration this should depress the proportion of marked ones recaptured equally along the whole column. But if the beetles have been dying during the experiment this is likely to depress the figures near the top of the column relatively more than those near the bottom because a larger proportion of the beetles marked early are likely to have died by December 3 than of those marked late. Consequently trends in columns may be taken as a measure of the death-rate in the population.

In Table 10.25*b* there is a clear trend along the rows suggesting that the population was being augmented by births throughout the period of the experiment and especially during the middle of the period. On the other hand the evidence from the columns does not indicate many deaths except perhaps towards the end of the period. The initial increase which we inferred from our estimates of N was associated with an early period when births were occurring in the virtual absence of deaths. Later the population declined because the birth-rate fell and the death-rate increased.

Instead of *Tribolium* the mealworm *Tenebrio molitor* may be used for this experiment. It has the advantage of being large and even an inexperienced person should have no difficulty in putting 9 or more distinctive marks on it. The disadvantage is that it develops so slowly. The modifications that are required include a larger box, larger sampling tubes and a bran-medium instead of a flour-medium.

10.3 THE MEASUREMENT OF DISPERSAL

The two experiments in this section are modelled on Dobzhansky and Wright's (1947) experiment with *Drosophila* (sec. 4.21). I have used *Paramoecium* for a laboratory model; larvae of *Tribolium* that are nearly fully grown dispersing in a flat box of flour are also quite good for a laboratory model. The snail *Helix* sp. has been used successfully for a field experiment. I have tried other species unsuccessfully. Most species that one might think of won't tolerate the intense crowding that they experience at the beginning of the experiment and still behave naturally. Yet if there are not a great many released in the centre at the beginning, the samples taken subsequently in the outer annuli are likely to contain too few animals. This difficulty is inherent in the design of this experiment. But I have not yet come across any other experiment that serves to measure the rate of dispersal quantitatively, as this one does.

Experiment 10.31

(*a*) Purpose

To measure the rate of dispersal of *Paramoecium*.

(*b*) Material

A culture of *Paramoecium* from which 30,000 may be collected and concentrated into about 10 to 15 c.c. of medium (Appendix I).

(*c*) Apparatus

(i) A flat dish with a glass bottom. We use one that is 60 cm. square and 10 cm. deep. For convenience the sides were made of galvanized iron and a sheet of glass was glazed in to make the bottom. But we found that we had to run a layer of beeswax around the edges to keep the medium away from the metal sides. Otherwise small quantities of metal were dissolved in the medium and killed the *Paramoecium*.

(ii) Medium made as in Appendix I, and strained through a sieve with about 15 mesh to the centimetre.

(iii) A sheet of paper a little larger than the floor of the dish marked with a pattern as shown in Figure 10.31*a*. The diameter of the central circle equals the width of the annuli and each annulus and the central circle are divided into sections of equal area.

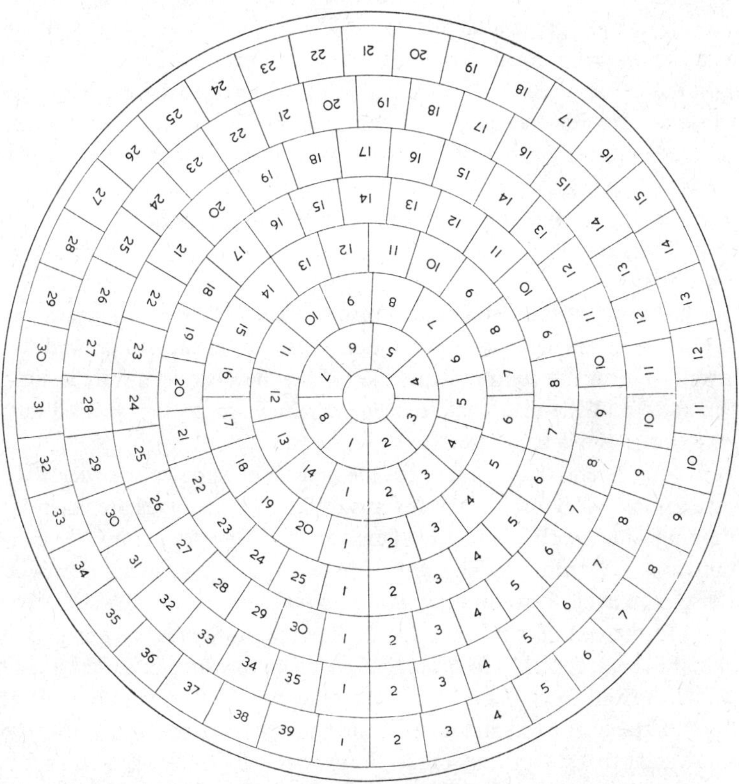

Fig. 10.31*a*. Design of an experiment to measure dispersal. The animals are let go in the central circle. Subsequently samples are taken in the central circle and in each annulus. The diameter of the central circle is equal to the width of an annulus. The annuli are divided into sectors of equal area as an aid to the placing of samples.

(iv) A sampling 'spoon'. We use a chemists' deflagrating spoon which has a capacity of about 1 c.c.

(v) Microscope slides with small cylinders, 15 mm. diam. by 4 mm. deep, cut from polythene tubing cemented on to them, with a waterproof cement. We find that two of these rings on one slide will hold a single sample of 1 c.c. Also these rings just fill the field of the 12·5 times objective of the binocular microscope which we find convenient for counting.

(vi) A table of random numbers.

(*d*) Method

(i) Medium is poured into the dish until it is about 2·5 cm. deep.

(ii) The *Paramoecium* may be collected and concentrated as follows. Several densely populated cultures of *Paramoecium* in glass jars about 9 cm. in diameter and 18 cm. deep are placed near a window in bright light but not in direct sunlight. After half an hour or so the *Paramoecium* will have congregated in the top half inch mostly near the edge. They can be drawn off with a fine pipette and collected in small dishes. By further pipetting they may be concentrated until there are about 30,000 in about 12 c.c. We have found this method satisfactory but there are several other ways of concentrating *Paramoecium* described by Wichterman (1953).

(iii) To release the *Paramoecium* place a glass or plastic cylinder about 2·5 cm. in diameter by 7 cm. long, open at both ends, and ground flat at one end, in the centre of the dish. Hold the cylinder firmly against the bottom of the dish while the *Paramoecium* are gently pipetted into it. Remove the cylinder gently. Note the time.

(iv) At the end of one hour take the first lot of samples and thereafter take samples at hourly intervals for 6 or 8 hours. Samples are taken by dipping a 'spoonful' of the medium out and tipping it into the containers for counting. Wash and wipe the spoon thoroughly between samples to avoid spreading the *Paramoecium* artificially – this is important. In the central circle and in the first annulus, especially at first, a spoonful may contain too many *Paramoecium*. So smaller samples may be taken from these places. We use a glass rod as in experiment 10.13. We calibrated the rod and found that the drop collected on the rod was equivalent to 1/25 of a spoonful. So in analysing the results we multiplied the average number per drop by 25 to make them equivalent to the average number per spoon.

(v) In order to estimate the density of the population in each annulus with the same precision it would be desirable to take about the same number of animals from each annulus (sec. 3.33). This would require more samples per unit area in the places where the animals were sparse than where they were dense, and in particular it would require a very large number of samples from the outer annuli in the early stages of the experiment. We did not do this but we compromised by taking samples in proportion to the area of the annulus and we did this by sampling from every third division in each annulus.

The set of divisions to be sampled in each annulus was chosen at random as follows. The divisions in the annuli were numbered as in Figure 10.31*a*. For each annulus and for each set of samples we selected a division at random by choosing a number from the table of random numbers. Starting from this division we worked round the annulus clockwise recording the number of every third division. Having thus worked out, in advance, which divisions from each annulus should be used we then arranged all of them (for the whole dish) in a random order, using the table of random numbers, and we sampled from them in this order. In this way we reduced the risk of introducing systematic errors which might have been important if we had sampled the annuli systematically, say from the centre outwards.

(*e*) Results

The results are given in Table 10.31*a*. The first step in estimating the rate of dispersal of the *Paramoecium* is to calculate the variance of the population as indicated by each hourly set of samples. The formula by which this may be done was derived in section 4.21. For example, for the third set of samples,

$$Sr^3f = 5{\cdot}33 \times 1 + 3{\cdot}60 \times 8 + 3{\cdot}29 \times 27 + \ldots + 0{\cdot}47 \times 512$$

and $$Srf + \frac{\bar{c}}{8} = 17{\cdot}00 \times 0{\cdot}125 + 5{\cdot}33 \times 1 + 3{\cdot}60 \times 2 + \ldots + 0{\cdot}47 \times 8 + \frac{17}{8}$$

The variances for each set of samples, calculated from the expression,

$$s^2 = \frac{Sr^3f}{Srf + \frac{\bar{c}}{8}}$$

are shown in Table 10.31*b*. They are first calculated in units of d^2 (d is the width of an annulus) and then converted to square centimetres. The standard deviation which is the square root of the variance is, of course, expressed in linear measure.

TABLE 10.31*a*

The number of each annulus is indicated by large Roman numerals. *Sf* stands for total *Paramoecium* taken in each annulus and $\bar{f}$ stands for the mean number per sample

No. of hours after release	*Number of* Paramoecium *in each annulus*																	
	Central circle		*I*		*II*		*III*		*IV*		*V*		*VI*		*VII*		*VIII*	
	Sf	$\bar{f}$	*Sf*	$\bar{f}$	*Sf*	$\bar{f}$	*Sf*	$\bar{f}$	*Sf*	$\bar{f}$	*Sf*	$\bar{f}$	*Sf*	$\bar{f}$	*Sf*	$\bar{f}$	*Sf*	$\bar{f}$
1	250	83·33	104	34·67	111	22·20	129	18·43	139	15·44	58	5·27	8	0·72	0	0·00	0	0·00
2	52	17·33	31	10·33	86	17·20	67	9·57	49	5·44	35	3·18	29	2·64	12	0·92	0	0·00
3	51	17·00	16	5·33	18	3·60	23	3·29	26	2·89	28	2·55	18	1·64	12	0·92	7	0·47
4	11	3·67	8	2·67	16	3·20	21	3·00	23	2·56	24	2·18	14	1·27	18	1·38	16	1·07
5	10	3·33	5	1·67	5	1·00	10	1·43	22	2·44	34	3·09	7	0·64	17	1·31	13	0·87
6	8	2·67	5	1·67	6	1·20	5	0·71	26	2·89	20	1·82	12	1·09	20	1·54	15	1·00
No. of samples	3		3		5		7		9		11		11		13		15	

TABLE 10.31*b*

Variances and standard deviations calculated from the data in Table 10.31*a*

No. of hours after release	*Variance in units of* d^2 $\dfrac{Sr^3\bar{f}}{Sr\bar{f} + \dfrac{\bar{c}}{8}}$	*Variance in sq. cm.*	*Hourly increment in variance in sq. cm.*	*Standard deviation (cm.)*
1	10·60	113·70		10·7
2	15·03	161·21	47·5	12·7
3	20·31	217·85	56·6	14·7
4	28·14	301·83	84·0	17·4
5	32·02	343·45	41·6	18·5
6	32·74	351·17	7·7	18·7

(*f*) Conclusions and comment

Up to the end of the fourth hour the variance increased fairly steadily. After the fourth hour the variance increased more slowly. This may have been because the *Paramoecium* were artificially restrained by the walls of the vessel in which the experiment was done. At least some of them had reached the walls by this time. So it may be best to base our conclusions on the first four hours.

If we care to assume that the distribution is sufficiently like a normal distribution we can make a number of inferences from the data in Table 10.31*b*.

(i) We know from the table of *t* (Fisher and Yates, 1948, Table III, bottom row) that half the population is likely to be no further from the centre than a distance equal to 0·6745 times the standard deviation. Four hours after the *Paramoecium* were released the standard deviation was 17·4 cm.; therefore half the population would be 11·7 cm. (0·6745 × 17·4) from the centre. At one hour, half the population was 7·2 cm. from the centre. Because, for a normal curve, the median coincides with the mean, this is equivalent to saying that at one hour the average distance of the *Paramoecium* from the centre was 7·2 cm. and at 4 hours the average distance was 11·7 cm.

(ii) Between the first and the fourth hours the variance increased on the average by 62·7 sq. cm. per hour. If we care to assume that, in the absence of artificial barriers, the variance would continue to increase at this rate, then we may predict that by the end of 24 hours the variance would be 1,555·8 sq. cm. (113·7 + 23 × 62·7). This gives a standard deviation of 39·4 cm. Multiplying this by 0·6745 we estimate that after 24 hours half the *Paramoecium* will be at least 26·6 cm. from the place where they were released.

(iii) The table of t shows that only 1% of the population in a normal distribution are further away from the centre than a distance equal to 2·5758 times the standard deviation. If we care to treat 1% as negligible then we may say that after 24 hours virtually all the *Paramoecium* will be within 101·5 cm. (39·4 × 2·5758) of the place where they were released.

Experiment 10.32

(*a*) Purpose

To measure the rate of dispersal of snails *Helix* sp. in the field.

(*b*) Material

A convenient number, between 500 and 1,000 snails.

(*c*) Apparatus

Eight-inch flower-pots to serve as 'traps' for the snails. In our experiment we used 119 pots. A tape measure. Some lacquer or paint suitable for marking the snails – we found that nail-polish stayed on best. And a table of random numbers.

(*d*) Method

(i) Choose a flat area of ground that is uniformly covered with grass or low herbage but not dissected with flower beds or shrubberies. A circular area about 60 feet in diameter may suffice unless the experiment is done at a time when the snails are unusually active.

(ii) Mark out a circle in the centre of the area 5 feet in diameter. Mark out 5 circles concentric with the central one having radii 7½, 12½, 17½, 22½ and 27½ feet.

(iii) Place one flower-pot upside down in the central circle and arrange the remaining 118 pots around the circumferences of the 5 concentric circles, spacing them 4 feet apart (or thereabout); but choose the position of the first one in each circle at random as in experiment 10.31. Prop the rim of the flower-pots off the ground so that the snails may crawl into them.

(iv) Mark the snails with a spot of varnish and release them in the central circle. We used 600 snails.

(v) Next day, and thereafter at daily intervals for the next 7 to 10 days count the marked snails sheltering under each pot and release them near by; then move the pots to a fresh set of random, but equally spaced positions around the same series of circles.

(vi) The distances suggested above are suitable for *Helix aspersa* during the winter in Adelaide. With another species or in different weather the snails may be more active and it may be necessary to increase these dimensions.

(*e*) Results

The results are set out in Table 10.32*a*. The methods of analysis are the same as for experiment 10.31, so the details need not be repeated here.

(*f*) Conclusion and comment

The decrease in variance between the 4th and 5th days may be put down to errors of random sampling. On the 9th and 10th days rather few snails were caught suggesting that by this time many of them had walked right off the area. If we consider the records up to the end of the 8th day we notice a rather large variation in the daily increments of variance. At least some of this was probably due to the influence of weather on the activity of the snails. In a rather more elaborate experiment it might be worthwhile to measure this as Dobzhansky and Wright (1947) did for *Drosophila*. At the end of the 8th day s was 13·04 feet and we may estimate that half the snails were at least 13·04 × 0·6745 feet (= 8·8 feet) from the centre at this time. Between the 1st and the 8th day the variance had increased by 22·0 square feet per day. At this rate the standard deviation on the 30th day would be 24·4 feet, and we may predict that half the snails would be at least 24·4 × 0·6745 feet (= 16·5 feet) away from the centre 30 days after they had been released.

TABLE 10.32*a*

No. of days after release	*Mean number of snails per trap*						*Variance s^2 in sq. ft.*	*S.D. s (feet)*	*Increase in variance (i)*
	Central circle	*Annulus*							
		I	*II*	*III*	*IV*	*V*			
1	30	6·500	0·000	0·000	0·000	0·000	15·9	3·99	
2	3	3·500	1·125	0·000	0·000	0·000	51·0	7·14	35·1
3	6	5·375	1·250	0·250	0·000	0·000	59·0	7·68	8·0
4	5	2·875	1·188	0·167	0·065	0·000	79·3	8·91	20·3
5	15	3·625	1·688	0·125	0·098	0·000	69·4	8·33	−9·9
6	18	2·625	2·313	0·250	0·226	0·026	101·0	10·05	31·6
7	3	1·875	1·563	0·625	0·194	0·026	143·8	11·99	42·8
8	6	1·375	1·438	0·708	0·258	0·077	170·1	13·04	26·3
9	4	2·375	1·250	0·458	0·129	0·308	203·0	14·25	32·9
10	0	0·500	0·375	0·250	0·290	0·128	294·8	17·17	91·8

CHAPTER XI

Physiological Responses to Temperature

11.1 THE INFLUENCE OF TEMPERATURE ON SPEED OF DEVELOPMENT

See section 5.11.

Experiment 11.11

(*a*) Purpose

To measure the duration of the life-cycle of *Tribolium confusum* and *T. castaneum* at several temperatures and to compare the responses of the two species to changes in temperature. There are three null hypotheses being tested concurrently in this complex experiment. (i) That there is no difference in the average length of the life-cycle (of the two species taken together) at 25°, 27° and 34°C. (ii) That, taking the three temperatures together, there is no difference in the length of the life-cycles of *T. castaneum* and *T. confusum.* (iii) If '(i)' and '(ii)' are disproved, that there is no difference between the species with respect to the length of the life-cycle at one temperature relative to that at another temperature.

(*b*) Material

Cultures of *T. confusum* and *T. castaneum* from which about 50 adults may be collected.

(*c*) Apparatus

(i) Several incubators. We used 3 that kept the temperature constant to ±0·2°C at 25°, 27° and 34°C.

(ii) Glass specimen tubes (shell vials) about 15 mm. wide by 5 cm. deep.

(iii) Flour-medium (see Appendix I).

(iv) Sieves made from bolting silk or nylon net of about 12 and 24 mesh to the centimetre.

(*d*) Method

(i) Eggs of a known age may be got by placing about 50 beetles in about 50 gm. of flour-medium of about 12% water-content at 27°C and leaving them there for 24 hours. At the end of 24 hours, sieve first through the coarse mesh to remove the beetles and then through the fine mesh to collect the eggs.

(ii) Divide the eggs at random into 3 groups. Place each group of eggs in a glass vial with about 8 gm. of flour-medium and place the vials at 25°, 27° and 34°C, or any other group of temperatures in this range that are convenient. Variability may be reduced by keeping the medium at about 12% water-content. But this is not essential; in the experiment reported below we had the medium rather drier than this; consequently there was a wide range between the first and the last beetles to emerge, especially at the lower temperatures (Table 11.11*a*).

(iii) After about 20 days at 34°, 32 days at 27° and 37 days at 25°C begin looking for adults. Sieve the medium every day, record the number of adults and remove them.

(*e*) Results

The results are given in Table 11.11*a*. The mean duration of the life cycle for both species at three temperatures is given at the foot of the table. For both species the life-cycle was shortest and least variable at 34°C, and longest and more variable at 25°C. At each temperature the life-cycle of *T. castaneum* was shorter than that of *T. confusum*. The differences seem quite large and the trends consistent (except for *T. confusum* at 25° and 27°C). So there would seem to be little doubt o their significance. Nevertheless I shall set out the appropriate statistical test so that we may know what confidence to have in the results that we have got. This test is called the 'analysis of variance'.

This test consists essentially of partitioning the total variance into two components, (i) that which is associated with differences in temperature or between species and (ii) that which is associated with the natural variability between individual beetles receiving the same treatment. If the former exceeds the latter by more than a certain ratio we accept the differences associated with the treatments as 'significant'. Moreover, the influence of the treatments may be analysed into three components: (i) that due to differences between the same species at different temperatures, (ii) that due to differences between the two species at the same temperature, and (iii) that due to differential responses of the two

TABLE II.11*a*

Days for *T. confusum* and *T. castaneum* to develop from egg to adult

	T. confusum						*T. castaneum*					
	25°		27°		34°		25°		27°		34°	
	Days (x)	*Beetles* (f)	*Days* (x)	*Beetles* (f)	*Days* (x)	*Beetles* (f)	*Days* (x)	*Beetles* (f)	*Days* (x)	*Beetles* (f)	*Days* (x)	*Beetles* (f)
	48	2	46	3	27	1	41	1	36	1	24	3
	49	1	47	1	29	1	42	2	37	1	25	2
	51	5	48	1	31	5	43	5	38	1	26	3
	52	2	49	5	32	7	44	1	39	2	27	4
	53	4	50	1	33	10	46	5	40	1	30	1
	54	2	51	5	34	10	48	2	41	2	31	1
	55	5	52	2	35	4	49	5	42	5	32	1
	56	4	53	3	36	3	50	2	43	2		
	57	8	54	4	37	2	53	2	45	1		
	58	5	55	5	38	2			48	2		
	59	6	56	1	39	3						
	60	5	57	6	41	2						
	61	2	58	3	42	1						
	62	2	61	3	45	1						
	63	1	62	3								
	64	2	63	6								
	66	1	65	3								
			67	3								
Mean, $\bar{x}$	56·6		56·1		34·4		46·2		41·6		26·7	
Variance, s^2	16·5		35·7		11·2		12·3		10·6		6·2	

species to changes in temperature. The method and the theory of the analysis of variance are discussed by Moroney (1953, pp. 233, 371).

The differential responses of the two species to changes in temperature may be expressed either as absolute differences, e.g.:

$$(56{\cdot}6 - 34{\cdot}4) - (46{\cdot}2 - 26{\cdot}7) = 22{\cdot}2 - 19{\cdot}5 = 2{\cdot}7 \text{ days}$$

or as relative differences, e.g.:

$$\frac{56{\cdot}6}{34{\cdot}4} - \frac{46{\cdot}2}{26{\cdot}7} = 1{\cdot}65 - 1{\cdot}73 = -0{\cdot}08$$

According to the first way of looking at it the life-cycle of *T. confusum* increased by 22·2 days as the temperature changed from 34° to 25°C whereas the life-cycle of *T. castaneum* increased by 19·5 days for the same change in temperature. According to the second way of looking at it the life-cycle of *T. confusum* increased by 65% and the life-cycle of *T. castaneum* by 73% as the temperature changed from 34° to 25°C. Both statements are logical but the second is more realistic because it tells us more about the biology of the two species of *Tribolium*. Having decided which way to pose the question we are now in a position to enquire whether the difference between the two ratios, 0·08, is significantly different from zero.

Now it so happens that if the analysis of variance were carried out directly on the data as they are given in Table 11.11*a* the test would indicate the significance of the differential responses on the absolute scale of days (i.e. does 2·7 differ significantly from zero?); whereas if the data are first transformed to logarithms, as in Table 11.11*b*, the test

TABLE 11.11*b*

The data from Table 11.11*a* transformed to logarithms

	T. confusum						*T. castaneum*					
	25°		27°		34°		25°		27°		34°	
	Log. Days (x)	*Beetles* (f)	*Log. Days* (x)	*Beetles* (f)	*Log. Days* (x)	*Beetles* (f)	*Log. Days* (x)	*Beetles* (f)	*Log. Days* (x)	*Beetles* (f)	*Log. Days* (x)	*Beetles* (f)
	1·68	2	1·66	3	1·43	1	1·61	1	1·56	1	1·38	3
	1·69	1	1·67	1	1·46	1	1·62	2	1·57	1	1·40	2
	1·71	5	1·68	1	1·49	5	1·63	5	1·58	1	1·42	3
	1·72	6	1·69	5	1·51	7	1·64	1	1·59	2	1·43	4
	1·73	2	1·70	1	1·52	10	1·66	5	1·60	1	1·48	1
	1·74	5	1·71	5	1·53	10	1·68	2	1·61	2	1·49	1
	1·75	4	1·72	5	1·54	4	1·69	3	1·62	5	1·51	1
	1·76	13	1·73	4	1·56	3	1·70	2	1·63	2		
	1·77	6	1·74	5	1·57	2	1·72	2	1·65	1		
	1·78	5	1·75	1	1·58	2			1·68	2		
	1·79	4	1·76	9	1·59	3						
	1·80	1	1·79	6	1·61	2						
	1·81	2	1·80	6	1·62	1						
	1·82	1	1·81	3	1·65	1						
			1·83	3								
Sf	57		58		52		23		18		15	
Sfx	99·88		101·30		79·81		38·21		29·08		21·40	
Sfx^2	175·078		177·055		122·573		63·503		46·996		30·550	
Mean, $\bar{x}$	1·75		1·75		1·54		1·66		1·62		1·43	
Variance, s^2	0·00108		0·00227		0·00157		0·00112		0·00092		0·00138	

measures the significance of the differential responses on the relative scale (i.e. does 0·08 differ significantly from zero?), which is what we want to know. The tests for the significance of the 'main effects' due to temperature and species may be properly done in the transformed scale because the geometric mean is an efficient estimate of the arithmetic mean. Accordingly we do the whole analysis on the data after they have been transformed to logarithms.

In the present instance there is another good reason for using the logarithmic transformation. The bottom two rows in Table 11.11*a* show that the variance tends to be correlated with the mean. Moreover there is a wide range in the variances – the largest is nearly 6 times the smallest. In these circumstances it is not permissible to pool the variances as is required in the analysis of variance. But the transformation to logarithms has got over the difficulty as may be seen from the means and the variances that are set out at the foot of Table 11.11*b*.

The figures from Table 11.11*b* may now be used to construct Table 11.11*c* which is called an 'interaction table'.

TABLE 11.11*c*

	25°		*27°*		*34°*		*Total*	
	Sfx	*Sf*	*Sfx*	*Sf*	*Sfx*	*Sf*	*Sfx*	*Sf*
T. confusum	99·88	57	101·30	58	79·81	52	280·99	167
T. castaneum	38·21	23	29·08	18	21·40	15	88·69	56
Total	138·09	80	130·38	76	101·21	67	369·68	223

The analysis then proceeds as follows:

$$\text{Correction factor (C.F.)} = \frac{(Sx)^2}{n}$$

$$= \frac{369{\cdot}68^2}{223}$$

$$= 612{\cdot}840$$

Total sum of squares (from figures in the body of Table 11.11*b*)

$$= Sx^2 - \text{C.F.}$$

$$= 175{\cdot}078 + 177{\cdot}055 + \ldots + 30{\cdot}550 - 612{\cdot}840$$

$$= 2{\cdot}915$$

Sum of squares for treatments (from figures in the body of Table 11.11*c*)

$$= S\frac{(Sx)^2}{n} - \text{C.F.}$$

$$= \frac{99{\cdot}88^2}{57} + \frac{101{\cdot}30^2}{58} + \ldots + \frac{21{\cdot}40^2}{15} - \text{C.F.}$$

$$= 2{\cdot}586$$

Sum of squares for temperature (from totals of colums in Table 11.11*c*)

$$= S\frac{(Sx)^2}{n} - \text{C.F.}$$

$$= \frac{138{\cdot}09^2}{80} + \frac{130{\cdot}38^2}{76} + \frac{101{\cdot}21^2}{67} - \text{C.F.}$$

$$= 2{\cdot}079$$

Sum of squares for species (from totals of rows in Table 11.11*c*)

$$= S\frac{(Sx)^2}{n} - \text{C.F.}$$

$$= \frac{280{\cdot}99^2}{167} + \frac{88{\cdot}69^2}{56} - \text{C.F.}$$

$$= 0{\cdot}410$$

Sum of squares for interaction (species × temperature) is got by subtracting the sum of squares for species plus the sum of squares for temperature from the sum of squares for treatments. Thus:

Sum of squares for interaction $= 2{\cdot}586 - (2{\cdot}079 + 0{\cdot}410)$
$= 0{\cdot}097$

The residual sum of squares can be got by subtracting the sum of squares for treatment from the total sum of squares. Thus:

Residual sum of squares $= 2{\cdot}915 - 2{\cdot}586$
$= 0{\cdot}329$

This last item may be cross-checked by summing the variances for the individual treatments. Thus:

$$0{\cdot}00108 \times 56 + 0{\cdot}00227 \times 57 + \ldots + 0{\cdot}00138 \times 14 = 0{\cdot}329$$

The remainder of the analysis of variance may be carried out in Table 11.11*d*. The variance ratio is got by dividing the appropriate mean square by the residual mean square. The probability is got from Table V of Fisher and Yates (1948). All the values are outside the scope of the table and we conclude that both the 'main effects' and the 'interaction'

TABLE 11.11*d*

Analysis of variance

Source of variance	*Degrees of freedom*	*Sums of squares*	*Mean square*	*Variance ratio*	*P*
Total	222	2·915			
Species	1	0·410	0·410	270	<0·000
Temperature	2	2·079	1·040	684	<0·000
Interaction (species × temp.)	2	0·097	0·049	32	<0·000
All treatments	5	2·586	0·517	270	<0·000
Residual	217	0·329			

are highly significant. Indeed variance ratios as large as these would occur by chance less often than once in 10,000 trials.

(*f*) Conclusions

All three null hypotheses were disproved at long odds. So we conclude: (i) That the life-cycle of *T. castaneum* is shorter than that of *T. confusum*. (ii) That for both species the life-cycle is shorter at higher temperatures. (iii) That this trend is more pronounced with *T. castaneum* than with *T. confusum*. If we had used 5 or 6 temperatures instead of 3 we may have been able to analyse the relationship between temperature and speed of development more closely. The next experiment with *Thrips imaginis* indicates how this may have been done.

Experiment 11.12

(*a*) Purpose

To measure the influence of temperature on the rate of egg-production by *Thrips imaginis*. There are two null hypotheses implicit in the design of this experiment. (i) That there is no difference in the number of eggs laid per day by *Thrips imaginis* at the different temperatures. (ii) That the daily number of eggs observed were no different from the numbers that would have been predicted, by drawing a straight line through the observations by the method of 'linear regression'. I have not troubled to calculate the precise odds against the first one because the significance of the results is obvious.

(*b*) Material

Thrips imaginis is indigenous to Australia. But this experiment could be done with any suitable local species of flower-inhabiting thrips

(Terebrantia). It is sufficient to breed or collect from the field about 60 to 100 females. Males will probably not be needed because most species of thrips lay eggs quite readily without being fertilized.

(*c*) Apparatus

(i) Several incubators that will keep the temperature constant within the range of about 13° to 25°C. We used five set at 10°, 13°, 16°, 20° and 23°C.

(ii) A number of small glass tubes about 5 mm. by 50 mm.

(iii) A fine camel-hair brush for handling the thrips.

(iv) Some stamens of *Antirrhinum, Penstemon* or some other fleshy stamen suitable as food for the thrips and for them to lay their eggs into.

(v) Microscope slides, gum chloral mounting medium (Berleses fluid or its equivalent) and a microscope for counting the eggs.

(*d*) Method

(i) Into each of, say, 100 tubes place one female thrips together with a fresh stamen. Take care to see that the anther is attached to the stamen because the thrips may not lay eggs without pollen. The thrips may be lightly anaesthetized with CO_2 if they are difficult to handle or to sex without anaesthetic.

(ii) Divide the hundred tubes at random into five lots, and place each lot of tubes in an incubator in a closed jar (or desiccator) over a saturated solution of NaCl. This will maintain the relative humidity at about 78%. We found the temperatures 10°, 13°, 16°, 20° and 23°C suitable for *Thrips imaginis* but other species may require a different range of temperature.

(iii) Every day for 10 to 15 days remove the old stamen and replace it with a fresh one, taking care not to harm or to lose the thrips.

(iv) Mount the old stamen under a coverslip on a microscope slide. Heat the mount gently over a small flame until the medium nearly boils. This will clear the plant-tissue but leave the eggs slightly opaque so that they can be easily counted under a microscope.

(*e*) Results

The results of our experiment with *Thrips imaginis* are set out in Table 11.12*a*. We kept the experiment going for 10 days and the table shows the average daily production of eggs for each of the thrips that were still alive at the end of the 10 days. That is why there are unequal numbers of thrips at different temperatures.

TABLE 11.12*a*

The mean daily number of eggs laid by *Thrips imaginis* at different constant temperatures

Temp. °C	*No. of thrips*	*Mean daily number of eggs laid by each thrips during a 10-day period*										*Total*	*Mean*
10	10	1·2	0·9	1·6	1·4	1·5	0·8	1·4	1·1	1·1	1·0	12·0	1·20
13	8	2·5	1·9	2·0	2·2	1·7	3·1	2·1	2·3			17·8	2·23
16	10	5·8	4·1	3·6	4·2	4·8	5·4	5·3	3·9	6·2	4·7	48·0	4·80
20	9	7·9	7·1	6·2	8·1	8·7	8·3	9·1	10·5	8·7		74·6	8·29
23	8	8·7	7·6	8·9	6·6	8·7	11·2	8·5	11·0			71·2	8·90

The relationship between temperature and rate of egg-production may be seen in Figure 11.12*a*. As the temperature rose from 10° to 23°C the rate increased from 1·2 to 8·9 eggs per thrips per day. The average increment in daily egg-production for unit increase in temperature over this range may be estimated by calculating the coefficient for linear regression.

$$b = \frac{S(x - \bar{x})(y - \bar{y})}{S(x - \bar{x})^2}$$

where x stands for temperature and y for number of eggs.

If the calculation is to be done on a machine it is best to write $S(x - \bar{x})(y - \bar{y})$ as $Sxy - \frac{Sx.Sy}{n}$ and $S(x - \bar{x})^2$ as $Sx^2 - \frac{(Sx)^2}{n}$.

There are 45 pairs of values x, y and therefore $n = 45$.

$$\begin{aligned} b &= \frac{4249{\cdot}000 - 3617{\cdot}351}{12744{\cdot}000 - 11777{\cdot}422} \\ &= \frac{631{\cdot}649}{966{\cdot}578} \\ &= 0{\cdot}6535 \text{ eggs per thrips per } 1^\circ\text{C} \end{aligned}$$

The theoretical relationship between temperature and daily rate of egg-production (assuming a steady increment in egg-production through the range 10°–23°C) may be expressed by the equation

$$\begin{aligned} Y &= \bar{y} + b(x - \bar{x}) \\ &= 4{\cdot}9689 + 0{\cdot}6535\,(x - 16{\cdot}1778) \\ &= -\,5{\cdot}6033 + 0{\cdot}6535x \end{aligned}$$

The straight line that corresponds to this equation is plotted in Figure 11.12*a*. The observed points do not follow the straight line very closely and the freehand curve (broken line) that has been drawn through them suggests that a sigmoid curve may have been better. This suggestion would be justified if we could show that the discrepancies between the observed points and the theoretical straight line are greater than might reasonably be expected by chance from errors of random sampling.

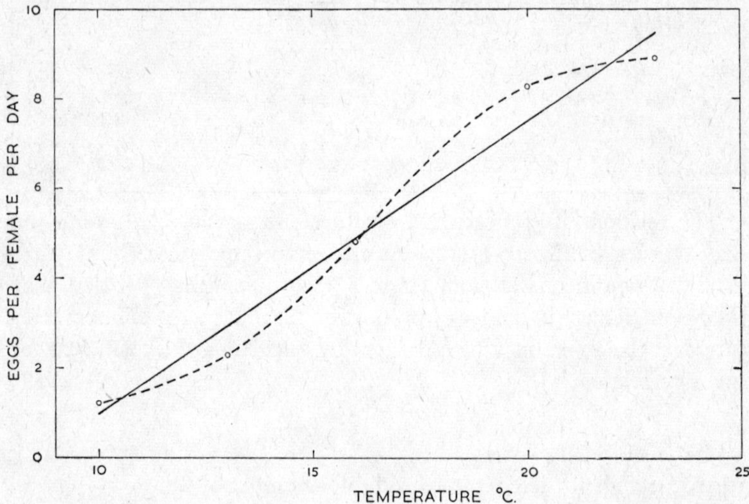

Fig. 11.12*a*. The influence of temperature on the rate of egg-production by *Thrips imaginis*.

This test is made by calculating χ^2 which is the ratio of two variances. In the numerator we put the variance that is due to systematic departures of the samples (as indicated by their means) away from the theoretical line. And in the denominator we put the variance due to the natural variability of the individual thrips in each sample.

Thus:

$$\chi^2 = S\frac{(Y - \bar{y})^2}{s^2}$$

where Y is the theoretical mean rate of egg-production as indicated by the straight line, $\bar{y}$ is the actual mean rate of egg-production as found by experiment and s^2 is the variance of $\bar{y}$

$$s^2 = \frac{S(y - \bar{y})^2}{n(n - 1)}$$

The calculation of χ^2 is set out (step by step) in Table 11.12*b*. There are 3 degrees of freedom for χ^2 and Table IV of Fisher and Yates

(1948) shows that the probability of such a large value for χ^2 arising from errors of random sampling is less than 1 in 1,000.

TABLE 11.12*b*

Calculation of χ^2 to test the goodness of fit of the straight line fitted to the data from Table 11.12*a*

Temp. °*C*	*Number of thrips*	Sy^2	$\frac{(Sy)^2}{n}$	$S(y-\bar{y})^2$	s^2	$(Y-\bar{y})^2$	$\frac{(Y-\bar{y})^2}{s^2}$
10	10	15·04	14·40	0·64	0·007	0·072	10·29
13	8	40·90	39·61	1·29	0·023	0·438	19·04
16	10	237·08	230·40	6·68	0·074	0·003	0·04
20	9	630·20	618·35	11·85	0·165	0·677	4·10
23	8	650·60	633·68	16·92	0·302	0·278	0·92

Total of last column $= \chi_3^2 = 34{\cdot}4$ $\quad P = <0{\cdot}001$

(*f*) Conclusion and comment

The highly significant value for χ^2 disproves the null hypothesis at long odds. We conclude that the straight line does not adequately represent the relationship between temperature and rate of egg-production. The trend seems to be sigmoid rather than linear and this suggests that the logistic equation may fit the data more closely (sec. 5.11). The method for calculating the constants of the logistic equation was described by Davidson (1944); and Browning (1952) described the method for testing the goodness of fit.

11.2 THE LETHAL INFLUENCE OF LOW TEMPERATURE

The method of 'biological assay' was first used to measure the responses of animals to drugs. It has now become the standard technique for testing insecticides and other poisons. It is the appropriate method for ecologists to use when measuring the 'tolerance' of animals to heat, cold, dryness and other harmful components of environment. The statistical method of 'probit analysis' which has been especially designed for analysing the results of such experiments has been discussed by Finney (1947). He described the theory of the method and worked through a number of examples which are very helpful to the beginner.

Experiment 11.21

(*a*) Purpose

To estimate the duration of exposure to 5°C that is required to kill half the population of *Calandra granaria*, which is also the best estimate that we can get of the mean length of life of *C. granaria* at 5°C (sec.

5.12). The null hypothesis is that there is no difference in the survival-time for *C. granaria* exposed to 5°C and those not so exposed but left at a favourable temperature (27°C in this experiment).

(*b*) Material

About 600 young adults of *Calandra granaria*. Weevils between the ages of 0 and 7 days may be got by sieving the adults out of a culture, leaving it in the incubator for 7 days and then sieving it again to collect the weevils that have emerged since the previous sieving.

(*c*) Apparatus

(i) Constant temperature cabinets set to run at 5°C and 27°C.

(ii) Six small screw-cap jars of about 20 c.c. capacity, with a ventilator of fine gauze built into the lid.

(iii) Some weevil-free wheat.

(*d*) Method

Count 100 weevils at random into each of the six jars and add a little weevil-free wheat. Place one jar, chosen at random from the six in the incubator at 27°C. Place the other five jars at 5°C. At intervals of 4, 7, 11, 14 and 18 days take one jar at random and place it at 27°C. After each jar has been at 27°C for 1 week, sieve the wheat and record the number of weevils dead and alive.

(*e*) Results

The results of a typical experiment are set out in Table 11.21*a*. This table and the analysis of the results are modelled on the example given by Finney (1947, p. 25 *et seq.*).

The first four columns in Table 11.21*a* are self-explanatory. The fifth column is derived from the fourth by correcting for the deaths that occurred in the controls. The correction is calculated by Abbot's formula as follows.

If p' is the total death-rate, c that part of the death-rate not associated with the treatment and p the death-rate caused by the treatment then,

$$p' = c + p(1 - c)$$

and

$$p = \frac{p' - c}{1 - c}$$

For present purposes the death-rate in the control may be taken as an adequate estimate of c.

The last two columns give the data from which Figure 11.21*a* was

TABLE 11.21*a*

The death-rate in samples of *Calandra granaria* exposed for various periods to 5°C

Days at 5°C	*No. of weevils* (*n*)	*No. dead* (*r*)	*Proportion dead* (*p′*)	*Estimated proportion killed by cold* (*p*)	*Log days* (*x*)	*Empirical probit*
0	95	6	0·063	0		
4	102	10	0·098	0·037	0·60	3·21
7	94	25	0·266	0·216	0·85	4·22
11	102	37	0·363	0·319	1·04	4·53
14	99	65	0·657	0·633	1·15	5·34
18	99	82	0·828	0·817	1·26	5·90

drawn. The 'dose' is transformed to logarithms because the probits fall more closely along a straight line when this is done. The probits are derived from the percentages given in column 5; they may be read from Table 1 of Finney (1947) or from Table IX of Fisher and Yates (1948).

The next step is to test the 'goodness of fit' of the provisional line

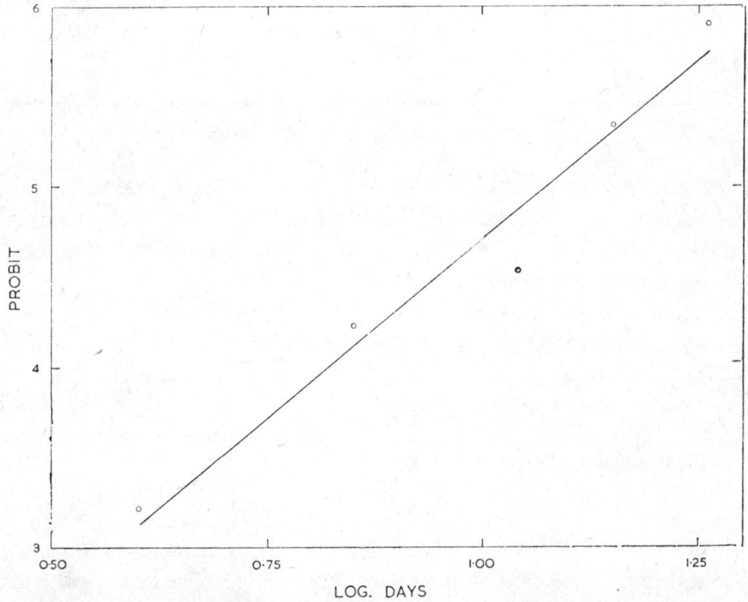

Fig. 11.21*a*. Probit-line for *Calandra granaria*. The abscissae are the logarithms of the number of days at 5°C.

that we have drawn in Figure 11.21*a* and we do this in Table 11.21*b*. The 'theoretical probits' are read from the graph and entered in Table 11.21*b* under the heading Y. In Table 11.21*b* x, n and r have the same meanings as in Table 11.21*a*; P is the proportion corresponding to Y, P' is derived from P by applying Abbot's formula, nP' is the number of deaths predicted by the graph after allowing for the deaths in the control, $r - nP'$ is the discrepancy between the observed and the predicted number of deaths; and the sum of the quantities in the last column gives an approximate estimate of χ^2. The transformation from P to P' is necessary because r includes deaths from causes other than the treatment and therefore the appropriate comparison is between r and nP' rather than between r and nP.

TABLE 11.21*b*

Test for 'goodness of fit' of the provisional probit line in Figure 11.21*a*

x	Y	P	P'	n	r	nP'	$r - nP'$	$\frac{(r - nP')^2}{nP'(1 - P')}$
0·60	3·12	0·03	0·091	102	10	9·3	0·7	0·06
0·85	4·12	0·19	0·241	94	25	22·7	2·3	0·32
1·04	4·85	0·44	0·485	102	37	48·5	11·5	5·15
1·15	5·30	0·62	0·641	99	65	63·5	1·5	0·10
1·26	5·74	0·77	0·784	99	82	77·6	4·4	1·14

$\chi_3^2 = 6{\cdot}77$ which is non-significant

In the present instance χ^2 is non-significant so we may complete our estimate of L.D. 50 and its reliability with the statistics provided by the provisional line of Figure 11.21*a*, without making any correction for heterogeneity. From the graph:

When $Y = 5$, $x = 1{\cdot}07$

Also, for unit increase in x, Y increases by 3·98.

So we can write,

$$m = 1{\cdot}07;\ b = 3{\cdot}98$$

Taking the antilog of m we get

$$\text{L.D. } 50 = 11{\cdot}7 \text{ days}$$

In order to discover the reliability of our estimate of m we must complete the calculations shown in Table 11.21*c*. The first 3 columns are copied from Table 11.21*b*; w is the weighting coefficient corresponding to Y which can be read from Table II of Finney (1947). The products

TABLE 11.21c

Estimate of the variance of m

x	n	Y	w	nw	nwx	nwx^2
0·60	102	3·12	0·159	16·22	9·732	5·839
0·85	94	4·12	0·476	44·74	38·029	32·325
1·04	102	4·85	0·635	64·77	67·361	70·055
1·15	99	5·30	0·616	60·98	70·127	80·646
1·26	99	5·74	0·517	51·18	64·487	81·254
Total				237·89	249·736	270·119

nw, nwx and nwx^2 are got by multiplying n, w, x, and x^2 together in the appropriate combinations.

The variance of m is given by

$$V_m = \frac{1}{b^2}\left(\frac{1}{Snw} + \frac{(m - \bar{x})^2}{Snw\,(x - \bar{x})^2}\right)$$

$$\bar{x} = \frac{Snwx}{Snw} = \frac{249\cdot736}{237\cdot90} = 1\cdot050$$

$$Snw\,(x - \bar{x})^2 = Snwx^2 - \frac{(Snwx)^2}{Snw}$$

$$= 270\cdot119 - \frac{(249\cdot736)^2}{237\cdot89}$$

$$= 7\cdot947$$

Therefore,

$$V_m = \frac{1}{3\cdot98^2}\left(\frac{1}{237\cdot89} + \frac{(1\cdot07 - 1\cdot05)^2}{7\cdot95}\right)$$

$$= 0\cdot000265 + 0\cdot000318$$

$$= 0\cdot000583$$

The standard deviation of m is given by

$$\text{S.D.}_m = \sqrt{V_m} = 0\cdot0241$$

The fiducial limits of m are given by,

$$\text{Fiducial limits of } m = m \pm t \times \text{S.D.}_m$$

For 5% probability $t = 1\cdot96$. Therefore,

$$\text{Fiducial limits} = 1\cdot07 \pm 1\cdot96 \times 0\cdot0241$$

$$= 1\cdot1172,\ 1\cdot0228$$

Taking antilogs we get,

Fiducial limits of L.D. 50 = 13·1, 10·5 days

(*f*) Conclusions and comment

We have estimated the L.D. 50 as 11·7 days. And the reliability of our estimate is such that if we were to repeat the experiment there is a 5% chance of getting an estimate that is either less than 10·5 or greater than 13·1 days. The L.D. 50 is the 'dose' of cold that is required to kill half the population. As was explained in section 5.12 the L.D. 50 may also be regarded as an estimate of the mean duration of life of *Calandra granaria* at 5°C.

We did not measure the survival-time of the controls at 27°C but since 94% of them were still alive at the end of the experiment the null hypothesis has clearly been disproved. But in this instance the disproving of the null hypothesis is less important than the estimation of the L.D. 50 and its fiducial limits.

This experiment was analysed by the approximate method of Finney. The graph, Figure 11.21*a*, is what Finney calls a 'provisional probit line' because it was merely drawn by eye through the provisional probits. For the simple problem of estimating the L.D. 50 and its fiducial limits the approximate method is often sufficient; and it saves a lot of work. Experiment 11.22 is more subtle and we shall have to use a more precise method of analysis, which is equally straightforward but more laborious.

Experiment 11.22

(*a*) Purpose

To estimate the duration of exposure to 5°C that is required to kill half the population of *Tribolium confusum* and *T. castaneum* in the pupal stage. And to test the null hypothesis 'that there is no difference between the L.D. 50 at 5°C for the two species'. The null hypothesis that is implicit in the first part of the experiment is that the survival-time for *Tribolium* that have been exposed to 5°C is no different from that of *Tribolium* that have been kept at a favourable temperature (27°C).

(*b*) Material

About 350 newly moulted pupae of each species.

(*c*) Apparatus

(i) Constant temperature cabinets running at 5° and 27°C.
(ii) Glass specimen tubes about 6 × 2·5 cm.

(iii) A desiccator or some other suitable glass vessel in which the relative humidity is maintained at about 70%.

(*d*) Method

(i) In order to get freshly moulted pupae collect rather more (perhaps twice as many) large, apparently fully fed larvae than the number of pupae that are required. Place them in flour-medium at 27°C. Each day collect the pupae from the medium and replace the larvae. The number of pupae collected each day should be divided equally, and at random, between all the treatments in the experiment. This should be repeated each day until the total number of pupae in each treatment approaches 50.

(ii) The pupae are placed in glass tubes with a little flour-medium. From each replicate one lot (the control) is placed directly at 27°C 70% R.H. The others are placed at 5°C to be removed after 2, 3, 5, 7, 9 and 11 days and placed with the controls at 27°C. They are left at 27°C until all the pupae have either emerged as adults or died. But they should be examined frequently and the adults removed as they appear. Otherwise they may eat those that are still pupae.

(*e*) Results

The results of a typical experiment are set out in Table 11.22*a*. The symbols have similar meanings to those in experiment 11.21.

Figure 11.22*a* is drawn from the data in columns 1 and 5. In this instance there is no need to transform days to logarithms because the relationship between empirical probits and days is already sufficiently linear.

One of the purposes of this experiment is to measure the significance of the difference between the L.D. 50 for the two species. We estimate the variance of the difference by pooling the variances of the estimates of L.D. 50 for both species. This implies that the two probit lines have the same slope. So we start by drawing the two provisional probit lines parallel. But before we can proceed to test the significance of the difference in L.D. 50 we must first test the lines to see if they really are parallel. If they differ significantly in slope then the device of drawing them parallel would result in an underestimation of the variance of the L.D. 50 and it would be necessary to increase the estimated variance by a 'heterogeneity factor'. For this test the approximate method of experiment 11.21 is not sufficient; it is necessary to calculate the probit regression equation by the method of 'maximum likelihood'. In practice

TABLE 11.22*a*

The death-rate in *Tribolium confusum* and *T. castaneum* exposed to 5°C

x	n	r	p ($c=0$)	Emp. probit	Y	y	w	nw	nwx	nwy
T. confusum										
0	39	0	0							
3	39	3	0·08	3·54	3·45	3·60	0·353	13·767	41·301	49·561
5	39	11	0·28	4·43	4·25	4·43	0·518	20·202	101·010	89·495
7	39	21	0·54	5·09	5·04	5·10	0·636	24·804	173·628	126·500
9	39	27	0·69	5·50	5·82	5·46	0·497	19·383	174·447	105·831
Total								78·156	490·386	371·387
T. castaneum										
			($c=0·04$)							
0	25	1	0							
3·5	25	5	0·17	4·05	4·27	4·06	0·444	11·100	38·850	45·066
5·5	25	12	0·46	4·90	5·08	4·90	0·589	14·725	80·988	72·153
7	25	20	0·79	5·81	5·67	5·80	0·511	12·775	89·425	74·095
10	25	24	0·96	6·75	6·30	6·63	0·321	8·025	80·250	53·208
Total								46·625	289·513	244·522

x = the duration of exposure to 5° in days
n = the number of pupae in each treatment
r = the number that died as pupae
p = the proportion dead, has been corrected by Abbot's formula for *T. castaneum* but not for *T. confusum* because none died in the control
Y = the theoretical probit is read from the graph in Figure 11.22*a*
y = the working probit is taken from Table IV of Finney (1947)
w = the weighting coefficient is taken from Table II of Finney (1947)
nw, nwx and nwy are got by multiplying n, w, x and y together in the combinations indicated by the headings to the columns.

this consists of starting with the provisional line, drawn by eye as before, and using the statistics from this line to calculate the regression coefficient by the maximum likelihood equations.

Therefore we draw, through each set of empirical probits in Figure 11.22*a*, what we judge by eye to be the best fitting straight lines subject only to the restriction that they must be parallel. Reading from these lines we may write the equations:

For *T. confusum* $Y = 2·27 + 0·39x$
For *T. castaneum* $Y = 2·91 + 0·39x$

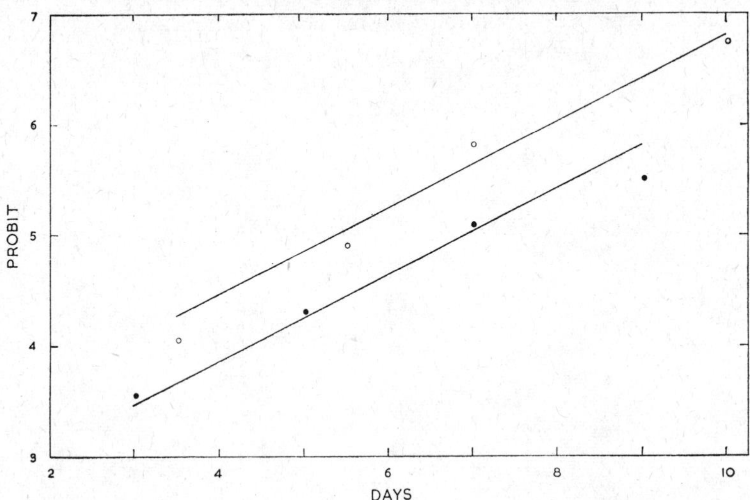

Fig. 11.22*a*. Probit-lines for *T. castaneum* (*hollow circles*) and *T. confusum* (*solid circles*). The abscissae are days at 5°C. The lines have arbitrarily been drawn parallel.

The values for Y in Table 11.22*a* may be calculated from these equations or read directly from the graph. From the data in Table 11.22*a* we calculate:

$Snwx^2$ by multiplying each value of $Snwx$ by x and summing the products

$Snwy^2$ by multiplying each value of $Snwy$ by y and summing the products

$Snwxy$ by multiplying each value of $Snwx$ by the corresponding value of y and summing the products

Then

$$Snw\,(x - \bar{x})^2 = Snwx^2 - \frac{(Snwx)^2}{Snw}$$

(written for convenience as S_{xx})

$$Snw\,(y - \bar{y})^2 = Snwy^2 - \frac{(Snwy)^2}{Snw}$$

(written as S_{yy})

$$Snw\,(x - \bar{x})(y - \bar{y}) = Snwxy - \frac{Snwx \times Snwy}{Snw}$$

(written as S_{xy})

The regression coefficient of probit on days, which measures the slope of the lines is given by:

$$b = \frac{S_{xy}}{S_{xx}}$$

In the present instance we have:

	T. confusum	*T. castaneum*	*Total*
$Snwx^2$	3414·372	2009·884	
$\frac{(Snwx)^2}{Snw}$	3076·903	1797·700	
S_{xx}	337·469	212·184	549·653
$Snwxy$	2434·137	1605·295	
$\frac{Snwx \times Snwy}{Snw}$	2330·250	1518·333	
S_{xy}	103·887	86·962	190·849
$\frac{Snwx}{Snw} = \bar{x}$	6·274	6·209	
$\frac{Snwy}{Snw} = \bar{y}$	4·752	5·244	

$$b \text{ (calculated from totals for both species)} = \frac{190{\cdot}849}{549{\cdot}653} = 0{\cdot}347$$

The lines with this slope pass through $\bar{x}$ and $\bar{y}$. Therefore we can write the equations:

T. confusum $\quad Y = 4{\cdot}752 + 0{\cdot}347(x - 6{\cdot}274)$
$\quad = 2{\cdot}575 + 0{\cdot}347x$

T. castaneum $\quad Y = 5{\cdot}244 + 0{\cdot}347(x - 6{\cdot}209)$
$\quad = 3{\cdot}089 + 0{\cdot}347x$

The difference between 0·39 and 0·35 (about 13%) in the two estimates of b is rather large, indicating that we did not judge the slope of the provisional lines (Fig. 11.22*a*) accurately enough. It will therefore be necessary to carry the maximum likelihood calculations through one more step in order to approach more closely to the best fitting lines. Table 11.22*b* is drawn up using the values of Y calculated from the equations given at the foot of the previous paragraph. The subsequent steps follow exactly as for Table 11.22*a*. From the statistics set out under the table we can calculate:

$$b = \frac{180{\cdot}053}{513{\cdot}518} = 0{\cdot}351$$

This value differs from the previous one by little more than 1%; so there is no need to carry the approximations any further.

TABLE 11.22*b*

Second approximation by the method of maximum likelihood; Y calculated from the equations $Y = 2{\cdot}575 + 0{\cdot}347x$, and $Y = 3{\cdot}089 + 0{\cdot}347x$

x	n	p	Y	$w(c = 0)$	nw	nwx	y	nwy
T. confusum								
3	39	0·08	3·62	0·309	12·051	36·153	3·60	43·384
5	39	0·28	4·31	0·535	20·865	104·325	4·42	92·223
7	39	0·54	5·00	0·637	24·843	173·901	5·10	126·699
9	39	0·69	5·70	0·532	20·748	186·732	5·48	113·699
Total					78·507	501·111		376·005
						$\bar{x} = 6{\cdot}383$		$\bar{y} = 4{\cdot}789$
T. castaneum				$(c = 4)$				
3·5	25	0·17	4·30	0·454	11·350	39·725	4·07	46·195
5·5	25	0·46	5·00	0·588	14·700	80·850	4·90	72·030
7	25	0·79	5·52	0·544	13·600	95·200	5·78	78·608
10	25	0·96	6·56	0·240	6·000	60·000	6·72	40·320
Total					45·650	275·775		237·153
						$\bar{x} = 6{\cdot}041$		$\bar{y} = 5{\cdot}195$

	T. confusum	*T. castaneum*	Total
$Snwx^2$	3527·979	1850·113	
$\frac{(Snwx)^2}{Snw}$	3198·597	1665·977	
S_{xx}	329·382	184·136	513·518
$Snwy^2$	1833·043	1266·265	
$\frac{(Snwy)^2}{Snw}$	1800·855	1232·016	
S_{yy}	32·188	34·249	66·437
$Snwxy$	2501·454	1511·302	
$\frac{Snwx \times Snwy}{Snw}$	2400·044	1432·659	
S_{xy}	101·410	78·643	180·053

The lines may now be tested to see whether they differ significantly from parallel. This is done by calculating a series of χ^2 from the sums of squares and products set out under Table 11.22*b*:

$$\chi_5^2 = SS_{yy} - \frac{(SS_{xy})^2}{SS_{xx}} \qquad \text{(Total)}$$

$$= 66{\cdot}437 - \frac{180{\cdot}005^2}{513{\cdot}518}$$

$$= 3{\cdot}31$$

$$\chi_2^2 = S_{yy} - \frac{(S_{xy})^2}{S_{xx}} \qquad (\textit{T. confusum})$$

$$= 32{\cdot}188 - \frac{101{\cdot}40^2}{329{\cdot}382}$$

$$= 0{\cdot}97$$

$$\chi_2^2 = 34{\cdot}249 - \frac{78{\cdot}643^2}{184{\cdot}136} \qquad (\textit{T. castaneum})$$

$$= 0{\cdot}66$$

The sum of the values for the individual species (i.e. 0·97 + 0·66 = 1·63) is subtracted from the total to give a χ^2 with 1 degree of freedom which tests the 'parallelism' of the lines (Table 11.22*c*).

TABLE 11.22*c*

	Degrees of freedom	χ^2
Departures from parallel	1	1·67
Residual heterogeneity	4	1·63
Total	5	3·31

None of these values for χ^2 is significant. So we may proceed to test the significance of the difference between the estimates of L.D. 50 for the two species without adjusting the data for heterogeneity.

The equations from the second approximation are:

T. confusum, $Y = 2{\cdot}549 + 0{\cdot}351x$

T. castaneum, $Y = 3{\cdot}075 + 0{\cdot}351x$

The value of x that gives $Y = 5$ is the estimated value of L.D. 50. For *T. confusum* L.D. 50 = 6·983 and for *T. castaneum* L.D. 50 = 5·484. The difference is 1·499. The variance of the difference is given by

$$V(d) = \frac{1}{b^2}\left[\frac{1}{S_1nw} + \frac{1}{S_2nw} + \frac{(\bar{x}_2 - \bar{x}_1 - d)^2}{SS_{xx}}\right] \quad \text{(See Finney, 1947, equation 5.7)}$$

$$= \frac{1}{9{\cdot}6221} + \frac{1}{5{\cdot}6241} + \frac{1{\cdot}3387}{63{\cdot}2659}$$

$$= 0{\cdot}3023$$

The standard deviation of the difference is the square root of the variance = 0·5498, whence we have

$$t = \frac{1{\cdot}499}{0{\cdot}5498}$$
$$= 2{\cdot}726$$

And $P < 0{\cdot}01$. (From Table II of Fisher and Yates, 1948.)

(*f*) Conclusions and comment

We have estimated that about 7 days at 5°C is sufficient to kill half the population of *T. confusum* and 5½ days' exposure is sufficient for *T. castaneum*. This is equivalent to saying that, on the average, *T. confusum* can live for about 7 days at 5°C and *T. castaneum* for 5½ days.

I mentioned this experiment in section 9.33. The following points serve to illustrate the argument that was developed in that section. The null hypothesis that there is no difference between the two species in regard to the time that they can stay alive at 5°C has been disproved with a probability of less than 1%. This is equivalent to saying that the odds against the two species being equally tolerant to exposure to 5°C are rather more than 100 to 1. No matter how often or how elaborately we repeated this experiment we would never be able to assert the inequality of the species with absolute certainty, but only with increasing degrees of probability. This is true of all scientific knowledge (sec. 9.33).

Compare the results of experiments 11.11, 12.11 and 11.22. In experiment 11.11 we demonstrated that the life-cycle of *T. castaneum* was reduced relatively more with increasing temperature than that of *T. confusum* (Table 11.11*d*, significant interaction between species and temperature). In experiment 12.11 we demonstrated that *T. castaneum* was more strongly influenced by 8 days' starvation and desiccation than *T. confusum* (Table 12.11*d*, significant heterogeneity between species after 8 days without food) and in experiment 11.22 we have shown that *T. castaneum* is significantly less 'cold-hardy' than *T. confusum*. These three facts are consistent with the statement that *T. castaneum* is adapted to living at a higher range of temperature than *T. confusum* and is, on the whole, less 'hardy' than *T. confusum*. This statement, strictly limited though it is to the summary of three facts, might be regarded as the beginnings of a *theory*. I explained in section 9.33 that all scientific theory is built up like this by finding generalizations that are consistent with the empirical facts. One can imagine how this theory, even at this early stage in its development, may give rise to speculation and argument leading to hypotheses and experiments. For example, do other species in this and related genera fit into a graded series according to their 'hardiness' and might this be related to their evolutionary origin – in the tropics or elsewhere? This is the way that scientific knowledge is built up, by proceeding through successive cycles of theory, model, hypothesis, null hypothesis and experiment to theory and so on.

The technique of biological assay and the associated statistical method of probit analysis are part of the essential techniques of ecology. The statistics may seem a little complex at first but with the help of Finney's (1948) useful account of the theory and especially the example that he works out in his Appendix I, the method can be mastered, for practical purposes, fairly easily.

I give below, in Table 11.22*d*, the results of an experiment with *Calandra oryzae* (large 'strain'). These results may be compared with those for *Calandra granaria* in experiment 11.21. It would be a useful exercise to estimate the L.D. 50 for *Calandra oryzae* and to determine whether it differed significantly from that for *C. granaria*.

TABLE 11.22*d*

The death-rate in *Calandra oryzae* (large 'strain') exposed to 5°C

No. of days at 5°C	*No. of weevils*	*No. dead*
0	85	1
4	99	2
7	88	29
11	97	76
14	101	92
18	106	101

CHAPTER XII

Behaviour in Relation to Moisture, Food and Other Animals

12.1 MOISTURE

See section 5.21.

Experiment 12.11

(*a*) Purpose

To observe the behaviour of the flour beetles *Tribolium confusum* and *T. castaneum* in a gradient of moisture. And to discover (i) whether the beetles move into the moist or the dry end of the gradient, (ii) whether their behaviour changes after they have been desiccated and starved for a period, and (iii) whether the two species differ from each other in their behaviour in a gradient of moisture.

(*b*) Material

About 500 adults of each species.

(*c*) Apparatus

(i) 10 specimen tubes (about 25 × 70 mm.) with gauze covers.

(ii) A desiccator containing a saturated solution of KOH (or concentrated H_2SO_4), and another containing a saturated solution of NaCl or some other solution giving a relative humidity of about 75%.

(iii) A constant temperature cabinet set at 27°C.

(iv) Several 'humidity-chambers' made as follows. One half of a petri dish (lid or bottom) is placed inside another larger one, and the inner one is packed up, if necessary until its upper surface is in the same plane as that of the larger dish. A strip of cardboard is used to make a partition that stretches centrally across both dishes and is flush with the top of the dishes as in Figure 12.11*a*. On one side of the partition the dishes are packed with moist sawdust, on the other side with sawdust that has been oven-dried at 105°C and allowed to cool in a desiccator. A sheet of nylon gauze is placed over the dishes and then the corresponding half-dishes (if lids have been used for the lower half of the humidity-chamber then lids must also be used for the upper half) are inverted over the gauze. The result, as in Figure 12.11*a*, is a circular chamber

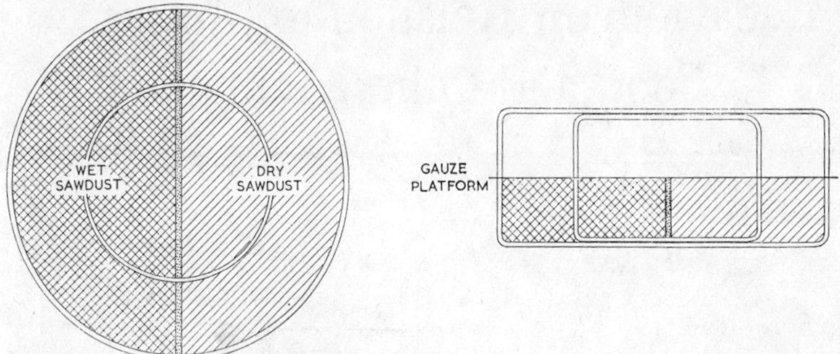

Fig. 12.11*a*. Humidity-chamber for measuring the response of *Tribolium* to a gradient in moisture. The beetles are placed in the inner compartment. The guard-ring serves to reduce the influence of secondary gradients which develop at the margin. The central partition comes flush with the top of the lower dish. The platform of nylon gauze on which the beetles move separates the lower dishes from the upper.

surrounded by a 'guard-ring'. One side of the chamber contains moist air above the moist sawdust and the other dry air. The existence of a gradient may be demonstrated by placing a strip of cobalt-thiocyanate paper across the chamber at right angles to the partition. The guard-ring reduces the chances of secondary gradients developing around the edges. To use the humidity-chambers the lids are lifted briefly, the beetles scattered over the gauze platform and the lids replaced immediately.

(*d*) Method

(i) Eight days before the experiment is to be done, collect 500 adults of each species. Divide each lot at random into 10 samples of 50 each. Place 2 lots of 50 into clean, dry tubes and put them in the desiccator at 15% relative humidity (saturated solution of KOH) at 27°C. Place the remaining 8 samples in tubes with a little flour-medium and put them at 70% relative humidity at 27°C.

(ii) At intervals of 2 days sieve out 2 lots of beetles from the flour-medium and place them in clean, dry tubes at 15% relative humidity. By the eighth day there will be 2 lots of 50 beetles that have been kept in the desiccator at 15% R.H. and 27°C without food for 2, 4, 6 and 8 days and 2 lots of 50 that have been in moist flour-medium all the time.

(iii) The next step is to set up the humidity-chambers and allow them to stand for 15 to 20 minutes to settle down. Then each sample of 50 beetles is tipped into a humidity-chamber and their behaviour watched and described. At intervals of one minute for upwards of 15 minutes the number on either side of the partition should be counted and recorded. The beetles respond best at temperatures above 25°C. If the temperature of the laboratory is below 25°C it may be necessary to do the experiment in a constant temperature room.

(*e*) Results

The result of an experiment in which *Tribolium castaneum* were kept without food for 0, 4 and 8 days before being tested in the humidity-chamber are set out in Table 12.11*a* and Figure 12.11*b*. The replicates agree fairly well so Figure 12.11*b* was drawn by combining the two replicates and transforming the totals to percentages.

TABLE 12.11*a*

The number of *Tribolium castaneum* on the dry side of the humidity-chamber

Minutes after release	*Days without food*					
	0 days		*4 days*		*8 days*	
	First replicate	*Second replicate*	*First replicate*	*Second replicate*	*First replicate*	*Second replicate*
1	20	33	15	18	6	1
3	38	36	22	20	6	3
5	38	35	20	12	2	2
7	34	30	22	20	5	2
9	35	35	20	23	2	3
11	37	33	21	22	3	3
13	38	29	20	19	3	3
15	30	33	15	18	4	1
20	30		20		5	
No. in replicate	50	50	50	50	21	23

From Table 12.11*a* we may interpolate to find the number of beetles on the dry side after 10 minutes, which may be taken as a measure of the beetles' response to the gradient of humidity. A similar experiment

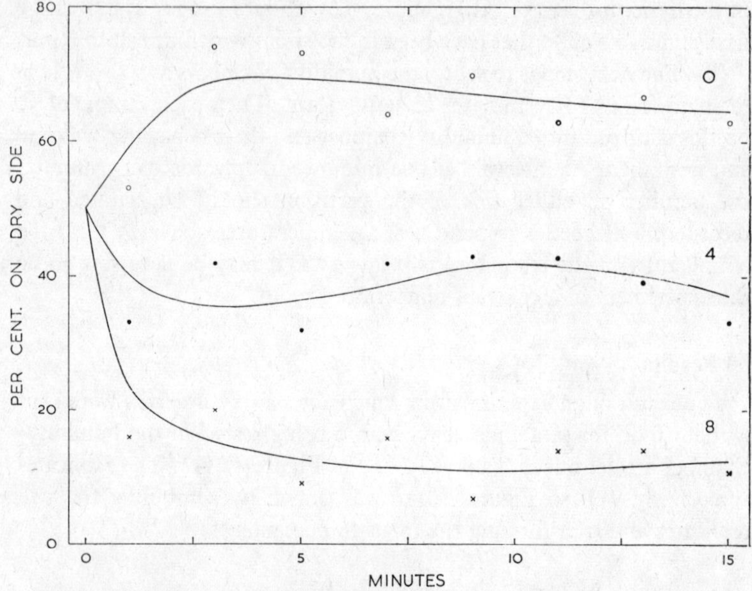

Fig. 12.11*b*. The number (expressed as a per cent) of *Tribolium castaneum* on the dry side of the humidity-chamber. The figures alongside the curves indicate the number of days of starvation before the experiment.

was done with *T. confusum*. The results for both species are shown in Table 12.11*b*.

About 70% of the beetles that had been living in the normal flour-medium up to the time of the experiment accumulated on the dry side of the humidity-chamber. But a majority of the beetles that had been without food for 8 days (87% of *T. castaneum* and 68% of *T. confusum*)

TABLE 12.11*b*

The behaviour of *Tribolium confusum* and *T. castaneum* in a gradient of moisture

Days without food	*T. confusum*			*T. castaneum*		
	No. on dry side	*Total*	*% on dry side*	*No. on dry side*	*Total*	*% on dry side*
0	74	98	76	70	100	70
4	42	100	42	43	100	43
8	24	78	31	6	44	13

were found on the wet side. These figures suggest (i) that the beetles recognize a gradient in moisture and tend to accumulate in one end or the other of it. (ii) That their response to the gradient depends on their condition – the normal tendency to accumulate on the dry side being reversed after 8 days' starvation. (iii) That this reversal of behaviour was more pronounced in *T. castaneum* than in *T. confusum*.

To measure the reliability of these conclusions we have to disprove the three null hypotheses (i) that the number of beetles on the dry side was not significantly different from the number on the wet side. (ii) That the distribution of the beetles did not depend on the number of days that they had been without food. (iii) That the distribution of *T. confusum* was not significantly different from the distribution of *T. castaneum*.

If the beetles do not recognize a gradient in humidity there should, on the average, be equal numbers of them on either side of the humidity-chamber and any discrepancy from the 1 : 1 ratio would have to be explained as errors of random sampling. By calculating a series of χ^2 as in Table 12.11*c* we can estimate the odds against the observed dis-

TABLE 12.11*c*

Species	*Days without food*	*No. of beetles*	*No. on the dry side*		χ^2 $\frac{2(O-E)^2}{E}$	*P*
			Observed	*Expected*		
T. confusum	0 4 8	98 100 78	74 42 24	49 50 39	25·5 2·6 11·5	<0·001 N.S. <0·001
T. castaneum	0 4 8	100 100 44	70 43 6	50 50 22	16·0 2·0 23·3	<0·001 N.S. <0·001
Total		520	259	260	0·004	N.S.

Total $\chi_6^2 = 80{\cdot}9$, $P<0{\cdot}001$

crepancies having arisen by chance. The sum of the 6 values of χ^2 in Table 12.11*c* is a χ^2 with 6 degrees of freedom. It is highly significant; there is less than one chance in a thousand that the observed discrepancies from the 1 : 1 ratio arose as errors of random sampling; apparently the beetles do recognize a gradient in moisture and distribute themselves unevenly in response to it. The difference between 80·9 and 0·004 gives a χ^2 with 5 degrees of freedom which measures the significance of the heterogeneity between the individual treatments. The heterogeneity is highly significant ($P < 0{\cdot}001$). An inspection of the table shows clearly that by far the greatest part of this heterogeneity is associated with the reversal of the behaviour of the beetles after they have been without food. Four of the six values for χ^2 are individually significant. The departures from the 1 : 1 ratio are significant for all the tests in the 0 and 8 days series. The fact that the results for 0 days are in the opposite direction from those for 8 days and that both are significantly different from the 1 : 1 ratio disproves ($P < 0{\cdot}001$) the second null hypothesis and establishes the conclusion that the response of the beetles to the gradient in moisture depends on the number of days that they have been without food.

But one wonders whether the difference between the species at 8 days (i.e. the relatively greater reversal in the behaviour of *T. castaneum*) may also contribute significantly to the heterogeneity between treatments. Because the analysis in Table 12.11*c* has established a significant discrepancy from the 1 : 1 ratio and a significant heterogeneity between days, the test for homogeneity between species must not assume either a 1 : 1 ratio or homogeneity between days. The correct procedure is to construct a 2 × 2 contingency table for each day separately as in Table 12.11*d*.

TABLE 12.11*d*

Test for homogeneity between species

Days without food		*T. confusum*		*T. castaneum*		*Total obs.*	χ^2 $S\frac{(O-E)^2}{E}$	*P*
		Obs.	*Exp.*	*Obs.*	*Exp.*			
0	Dry side	74	71·3	70	72·7	144		
	Wet side	24	26·7	30	27·3	54	0·74	N.S.
	Total	98	98	100	100	198		
4	Dry side	42	42·5	43	42·5	85		
	Wet side	58	57·5	57	57·5	115	0·02	N.S.
	Total	100	100	100	100	200		
8	Dry side	24	19·2	6	10·8	30		
	Wet side	54	58·8	38	33·2	92	4·42	<0·05
	Total	78	78	44	44	122		

There was significant heterogeneity between the species at 8 days ($\chi^2 = 4{\cdot}4$) but not at 0 or 4 days. The heterogeneity at 8 days was due to the relatively greater response by *T. castaneum* to the period of starvation (Table 12.11*b*).

(*f*) Conclusions and comment

Beetles of both species in their normal freshly fed condition moved towards the dry side of the gradient and tended to stay there. Beetles that had been starved for 8 days moved towards the wet side of the gradient. There was no significant difference in the behaviour of the two species in the 0- and 4-day lots. About half the beetles that had been without food for 4 days were found on either side of the humidity-chamber. Perhaps half the beetles retained their original preference for dry air while the other half had come to prefer moist air. Or, perhaps all the beetles lost their capacity to recognize a gradient in moisture. This experiment does not allow us to choose between these two explanations. The difference between the species after 8 days without food suggests that *T. confusum* may be 'hardier' than *T. castaneum.*

12.2 FOOD

Experiment 12.21

(*a*) Purpose

To discover whether the females of the white cabbage butterfly *Pieris rapae* are more likely to lay their eggs on Brassicae than on other sorts of plants that occur commonly in gardens where Brassicae may be grown.

(*b*) Material

About 30 adults of *Pieris rapae.* We usually rely on catching them in the field rather than take the trouble to maintain a laboratory stock.

(*c*) Apparatus

(i) About 30, 6-inch flower-pots and suitable potting soil.

(ii) Six seedlings of cabbage or some other Brassicae, *Nasturtium, Geranium, Sonchus* and lettuce, or any other similar group of plants. Also some seeds or seedlings of *Alyssum, Erigeron* or some other sort of plant that will grow well in pots and provide plenty of flowers with nectar for the butterflies to feed on.

(iii) A cage of fly-wire gauze about $2 \times 1 \times 1$ m. A convenient cage has a trap-door lid through which the pots may be arranged, and cloth sleeves in the sides which allow access to the cage after the butterflies have been put in.

(*d*) Method

(i) Some weeks before the experiment is to be done the plants should be established in the pots. It may take longer to get the flowering plants going well.

(ii) When the experiment is to be done place several pots of flowering plants at one end of the cage and arrange the plants for egg-laying at the other end. Use 5 of each sort and arrange the 25 pots in the form of a randomized block. In this instance the randomized block has 5 rows with 5 plants in each row. Each row contains one and only one of each sort of plant. And within each row the positions of the different sorts of plants occur at random. This arrangement is best attained by numbering the pots and then arranging them in the rows according to numbers drawn from a table of random numbers. The arrangement that was used in the present experiment is shown in Table 12.21*a*.

TABLE 12.21*a*

						Key
I	A	C	E	D	B	A *Nasturtium*
II	E	D	A	B	C	B *Geranium*
III	A	C	B	D	E	C Brassicae
IV	E	B	A	D	C	D Lettuce
V	A	B	D	E	C	E *Sonchus*
		Alyssum				

(iii) After the plants have been arranged catch about 30 butterflies and let them go in the cage. Leave them there for several days. This experiment is best done in the open air or in a glass house in sunny weather, because the butterflies lay eggs more freely when the sun is shining.

(iv) After several days take the plants out of the cage and count the eggs that have been laid on them. Take care not to dislodge the eggs. Keep a record of the position of each plant and the number of eggs on it as in Table 12.21*b*.

(*e*) Results

The results of a typical experiment are shown in Table 12.21*b*.

The differences in the mean number of eggs laid on the Brassicae and on all the other sorts of plants are so great and so consistent that we may take them as highly significant without troubling to calculate precise probabilities. The extreme heterogeneity in the variances would, in any case, make the calculation of precise odds impracticable. The

TABLE 12.21*b*

The number of eggs laid by *Pieris rapae* on 5 different sorts of plants, arranged in a randomized block as indicated in Table 12.21*a*

						Total		
I	12	42	0	0	0	54	*Nasturtium*	51
II	0	0	12	3	24	39	*Geranium*	4
III	6	81	1	0	0	88	Brassicae	359
IV	0	0	13	0	105	118	Lettuce	0
V	8	0	0	0	107	115	*Sonchus*	0
					Total	414	Total	414

difference between the number of eggs on *Nasturtium* and *Geranium* also seems to be quite beyond doubt but in this case one may, if one wishes, calculate the precise significance of the difference between the means because the variances seem to be reasonably homogeneous. The variance of the difference between the means is given by:

$$s^2 = \frac{1}{n_1 + n_2 - 2}[S(x_1 - \bar{x}_1)^2 + S(x_2 - \bar{x}_2)^2]$$

For the significance of the difference between means we require to estimate

$$t = \frac{\bar{x}_1 - \bar{x}_2}{s} \times \sqrt{\frac{n_1 \times n_2}{n_1 + n_2}}$$

In the present instance:

$$s^2 = \frac{1}{8}(6{\cdot}8 + 6{\cdot}8)$$

$$s = 1{\cdot}304$$

$$t = \frac{10{\cdot}2 - 0{\cdot}8}{1{\cdot}304} \times \sqrt{2{\cdot}5}$$

$$= 11{\cdot}4$$

$$P < 0{\cdot}001$$

(*f*) Conclusion and comment

The tendency for *Pieris* to lay most of its eggs on Brassicae in preference to *Nasturtium, Geranium, Sonchus* or lettuce has been established at a very high though unknown level of probability. The tendency to lay more eggs on *Nasturtium* than on *Geranium* was established with a probability of less than 0·001.

The heterogeneity of the variances over the whole experiment precluded a full analysis of the randomized block which would have given an estimate of the significance of the difference between rows. But there is just a suggestion that the rows near the *Alyssum* are likely to have more eggs laid on them than the others.

It so happened that, in this experiment, the 5 Brassicae comprised one each of cabbage, broccoli, brussels sprout, cauliflower and greenleaf cauliflower. The number of eggs laid was greatest (107) on the broccoli and least (24) on the brussels sprout. Was this just an accident of random sampling or does it reflect a tendency that is likely to be repeated on another occasion? The present experiment gives no answer to these questions because the varieties of Brassicae were not replicated. So we design experiment 12.22.

Experiment 12.22

(*a*) Purpose

To discover whether *Pieris rapae* is likely to lay different numbers of eggs on different varieties of Brassicae when they are given equal opportunity to lay on any of them.

(*b*) Material

About 30 adult *Pieris rapae.*

(*c*) Apparatus

The same as for 12.21 except that the 25 plants for egg-laying comprise 5 varieties of Brassicae, e.g. broccoli, cabbage, brussels sprout, cauliflower and greenleaf cauliflower.

(*d*) Method

The same as for experiment 12.21 except that the plants are arranged in a latin square instead of a randomized block. In the latin square each variety is represented once and once only in each row and once and once only in each column. Apart from this restriction their positions in the rows and columns are random. This design allows gradients, if they exist, to be picked up across rows and across columns, and in addition enhances the precision of the comparison of varieties. In the present instance the comparisons between rows and columns are desirable because it seems likely that the presence of the *Alyssum* at one end of the cage may cause a gradient in the number of eggs laid; also vagaries of sun, shadow, wind, etc. may cause a gradient in another direction.

The method for selecting a latin square at random is given by Fisher and Yates (1948), Table XV. The design that was used in the present experiment is shown in Table 12.22*a*.

TABLE 12.22*a*

Design of latin square

	(*i*)	(*ii*)	(*iii*)	(*iv*)	(*v*)
I	B	D	E	A	C
II	C	E	A	B	D
III	D	C	B	E	A
IV	A	B	C	D	E
V	E	A	D	C	B

Alyssum

Key

A Greenleaf cauliflower
B Cauliflower
C Broccoli
D Cabbage
E Brussels sprout

(*e*) Results

The number of eggs on each plant are given in Table 12.22*b* in the positions in which they occurred in the experiment.

TABLE 12.22*b*

Number of eggs laid by *Pieris* on 5 varieties of Brassicae arranged in a latin square

	(*i*)	(*ii*)	(*iii*)	(*iv*)	(*v*)	*Total*	*Varieties*	*Eggs*
I	42	18	30	6	20	116	Greenleaf cauliflower	110
II	54	4	3	6	18	85	Cauliflower	83
III	4	5	2	12	53	76	Broccoli	122
IV	15	16	9	14	33	87	Cabbage	70
V	13	33	16	34	17	113	Brussels sprout	92
Total	128	76	60	72	141	477	Total	477

From an inspection of the table it may be seen that more eggs were laid on broccoli than on any other variety and also more eggs were laid round the edges than in the centre of the square. The differences are only moderate and it is rather doubtful whether they will be significantly greater than the variances within varieties. An analysis of variance will answer this question precisely. For the analysis of variance the sums of squares for total, rows, columns and varieties are calculated as follows,

$$\text{Sum of squares,}\quad S(x-\bar{x})^2 = \frac{Sx^2}{n'} - \text{C.F.}$$

For rows, x is the total for each individual row

$$n' = 5$$

For columns, x is the total for each individual column

$$n' = 5$$

For varieties, x is the total for each individual variety

$$n' = 5$$

For total, x is each individual figure in the body of the table

$$n' = 1$$

The correction factor C.F. is the same for all expressions,

$$\text{C.F.} = \frac{(Sx)^2}{n} = \frac{477^2}{25}$$

The sums of squares are set out in Table 12.22*c*.

TABLE 12.22*c*

	$\frac{Sx^2}{n'}$	*C.F.*	$S(x - \bar{x})^2$
Rows	9,359	9,101·16	257·84
Columns	10,165	9,101·16	1,063·84
Varieties	9,447·4	9,101·16	346·24
Total	14,529	9,101·16	5,427·84

The analysis of variance is given in Table 12.22*d*

TABLE 12.22*d*

Analysis of variance

Source of variance	*Degrees of freedom*	*Sums of squares*	*Mean square*	*Variance ratio*	*P*
Columns	4	1,063·84	265·96	<1	N.S.
Rows	4	257·84	64·46	<1	N.S.
Varieties	4	346·24	86·56	<1	N.S.
Residual	12	3,759·92	313·33		
Total	24	5,427·84			

None of the variance ratios is significant.

(*f*) Conclusion and comment

There is no evidence from this experiment that the distribution of the eggs depended on the position of the row or the column or on variety.

This is not to say that significant differences might not be demonstrable with a larger experiment. With a sufficiently large number of replicates even quite small differences may be shown to be significant.

12.3 OTHER ANIMALS OF THE SAME KIND

Experiment 12.31

(*a*) Purpose

To observe the outcome of crowding *Tribolium confusum* and *T. castaneum* in a limited volume of medium when the medium is renewed at regular and frequent intervals.

(*b*) Material

About 400 pairs of adult beetles of each species. Since they are more easily sexed as pupae than as adults it is best to begin with rather more than 400 pairs of pupae of each species.

(*c*) Apparatus

(i) Two constant temperature cabinets set to run at 25° and 34°C or some other convenient temperature in this range.
(ii) Seventy-two glass tubes about 5 × 1·5 cm.
(iii) Coarse and fine sieves having about 10 and 20 mesh per cm.
(iv) a supply of flour-medium.

(*d*) Method

(i) Put 8 gm. of flour-medium in each tube.
(ii) Distribute the beetles through the tubes in conformity with the plan shown in Table 12.31*a*. Label the tubes accurately and put them away at the appropriate temperatures.

TABLE 12.31*a*

Species	*T. confusum*						*T. castaneum*					
Temperature	25°C			34°C			25°C			34°C		
Density (in pairs)	2	10	20	2	10	20	2	10	20	2	10	20
Replicates	Six in every treatment											

(iii) Once every week sieve the culture through the coarse sieve; count the adults, pupae and large larvae that are held back by the sieve; return them to the tube. Then sieve the medium again with the fine sieve. Return the eggs and small larvae to the colony and throw away the old medium. Add 8 gm. of new medium.

(iv) Continue with this routine for as long as is practicable. Some interesting results may begin to emerge after about 18 to 20 weeks but the experiment will become more instructive if it is carried on for longer than this.

(*e*) Results

The history of the population in each of the 12 treatments may be illustrated by curves like those in Figure 12.31*a*. The curves are the means of all the replicates at each treatment. If the flattening of the

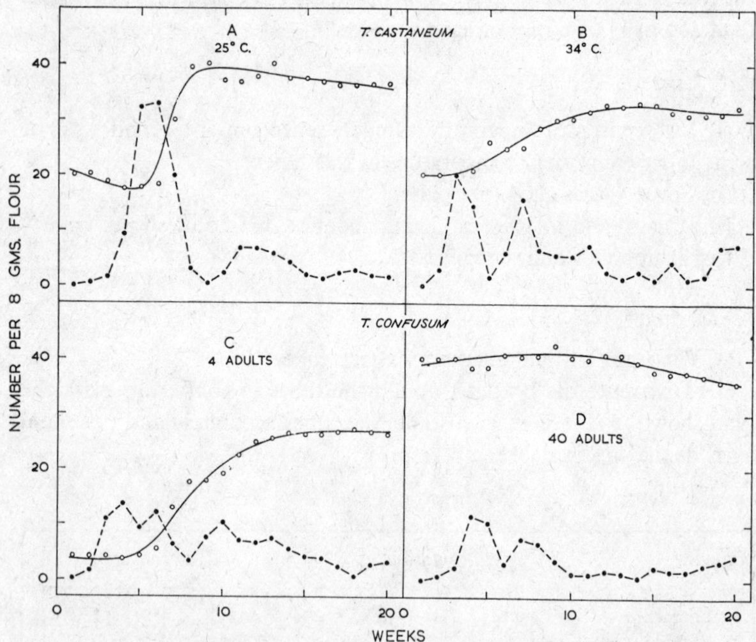

Fig. 12.31*a*. Populations of *T. castaneum* and *T. confusum* that started at different levels of crowding or were kept at different temperatures. *Solid lines*, adults; *broken lines*, large larvae and pupae. Both lots of *T. castaneum* started with 20 adults but were kept at different temperatures; both lots of *T. confusum* were kept at the same temperature but were started with different numbers of adults. For details of other treatments in this experiment see Table 12.31*b*.

curve for adults which is characteristic of the four curves in Figure 12.31*a* should occur in all 12 treatments then the results may be summarized as in Table 12.31*b*. The figures in the table are the mean number of adults in each replicate of each treatment for the 12th to the 20th week of the experiment. There are fewer than 6 replicates left in every treatment because some were lost through accidents during the course of the experiment.

(*f*) Conclusion and comment

The most striking feature of these results is that, in every replicate of every treatment, after about 10 weeks the population of adults had reached a density which seemed to be characteristic of the treatment or even of the replicate and then remained virtually steady for the remainder of the experiment. There was surprisingly little difference in the densities achieved after 20 weeks in the different treatments. There were rather more beetles in the treatments that had started with 20 or 40 beetles but the differences were not striking. The differences between replicates (Table 12.31*b*) are more striking. It seems as if this might be associated with genetical differences between replicates. An experiment might be designed to test this idea.

TABLE 12.31*b*

The mean number of adults of *Tribolium confusum* and *T. castaneum* in each replicate during the 12th to 20th weeks

Species	*Temperature °C*	*Density at beginning (pairs)*	*No. of adults in each replicate*					*Mean*
T. confusum	25°	2	28·0	30·8	33·2	9·0		25·3
		10	27·0	62·2	68·3	20·8	46·2	44·9
		20	32·0	42·7	51·3	36·0	38·9	40·2
	34°	2	30·2	19·9	23·9	31·6		26·4
		10	32·3	32·2	30·6			31·7
		20	29·2	35·3	46·9	40·7		38·0
T. castaneum	25°	2	6·4	7·8	9·1	9·4		8·2
		10	39·6	43·9	18·0	47·2		37·2
		20	56·2	38·7	58·8	43·6	57·9	51·0
	34°	2	37·1	34·6	20·4	20·6	22·6	27·0
		10	38·2	31·2	26·7			32·0
		20	61·1	54·3	45·4	39·9	37·4	47·6

The relatively low numbers of larvae and pupae and the few pupae that develop to the adult stage suggest that the numbers of adults fail to increase beyond a certain characteristic density because the adults, and perhaps the larvae, eat the immature stages especially the eggs and pupae. An experiment to discover the causes of death of *Tribolium* when they are forced to live in a crowd would therefore be worth doing.

Experiment 12.32

See section 6.122

(*a*) Purpose

To observe 'territorial behaviour' in wild house mice *Mus musculus*.

(*b*) Material

About 30 or 40 pairs of house mice. Mice that have recently been trapped in the wild are good. But mice that have been reared for several generations in the laboratory may also be used.

(*c*) Apparatus

A number of cages made and furnished as follows. The cage consists of a box about 1 m. cube open at the top. It is either made from, or lined with galvanized flat iron about 26 gauge. At the centre of each side just above the floor there is an exercise-wheel. In each corner and also along the walls there are nesting boxes – about 6 or 8 altogether. Each nesting box is about $18 \times 12 \times 8$ cm. It has a doorway at one end and a lid that can be opened to look at the mice. It contains strips of paper or other material suitable for making nests. In the centre of the cage there is a supply of water, and at several places around the walls supplies of food.

(*d*) Method

The mice are marked so that they may be recognized; toes may be clipped for this purpose. Then 4 males and 4 females are let go in each cage. The mice should be watched closely without disturbing them too much and a daily record made of their behaviour.

(*e*) Results

Nearly always the following events take place – with minor variations in detail. After a relatively short period all the females come to live together in one box usually with only one male. The males fight and quite soon one or more are killed or die; in due course there will be only one male left alive. Almost invariably it will be the one that was earlier observed to be living with the four females. Notwithstanding the fact that the family group is established quite early it is unusual for any of the females to become pregnant until the last of the 'inferior' males has been killed or has died.

This experiment, as I have described it, merely provides an opportunity to observe and describe one aspect of the 'territorial' behaviour of mice. But the method permits of extensions in a number of directions. For example, by matching mice of different genotypes, ecotypes or with different histories it is possible to compare the 'fighting success' of the different types.

12.4 A PLACE IN WHICH TO LIVE

Experiment 12.41

(*a*) Purpose

To discover the sorts of places where the isopods *Porcellio laevis* and *P. scaber* live and to recognize the more important components in their environments.

The students work for perhaps half a day in a garden that is chiefly shrubbery and orchard. There are several heaps of compost of different size, age and composition. There are small heaps of grass cuttings and part of the garden has been mulched with a layer of grass-cuttings taken from a lawn. There are a number of flat stones and boards lying around, some on the area that has been mulched and some on bare ground. Some parts of the area are watered by sprinklers regularly about once a week during summer and other parts are not watered at all or only rarely. The summer in Adelaide is arid, so there is a marked difference in moistness and in vegetation between the watered and unwatered areas. There are two species of isopods living in the garden.

There are also many blackbirds, mynahs, magpies, fantails and other insectivorous or omnivorous birds.

The students are asked to discover and describe the sorts of places where the two species of isopods live, and to form an opinion, so far as they can, on which components of environment are likely to be most important in determining the distribution and abundance of the isopods. And they are asked to draw up a skeleton plan for an investigation of the ecology of the isopods.

Moisture, food and predators are obviously important components of environment and there are a number of interactions between predators and a place in which to live, which are usually discussed at length,

And there is scope for speculation on how far the two species, apparently living together in the same sorts of places, influence each other's chance to survive and multiply.

The idea behind this exercise may be extended in a number of directions. In some localities certain animals that live in freshwater streams or ponds may be especially suitable for a similar exercise.

Appendix

Materials for Experiments

The following species have been used in the experiments described in Part II.

Paramoecium sp., *Helix aspersa*, *Porcellio* spp., *Thrips imaginis*, *Saissetia oleae*, *Calandra oryzae*, *Calandra granaria*, *Tribolium castaneum*, *Tribolium confusum*, *Tenebrio molitor*, *Pieris rapae*, *Mormoniella vitripennis* and *Mus musculus*.

We do not keep laboratory colonies of *Helix*, *Porcellio*, *Thrips*, *Saissetia*, *Pieris* or *Mus* because we find it easier to catch wild ones in the field when we need them. Laboratory colonies of the other species may be maintained as follows.

Paramoecium

Glass jars containing about 1 litre are suitable. The medium is made by boiling lawn-cuttings for about 30 minutes in tap-water, allowing it to cool, straining it and then leaving it on the bench for 5 days until it is heavily populated by bacteria. It is then inoculated with a few *Paramoecium*. At room temperatures the colony usually becomes quite dense in about 5 days and after about 7 days it requires to be sub-cultured.

Calandra oryzae (small) and *C. granaria*

Tin cans with press-in lids and a capacity of about 2 litres are quite good. Ventilators of fine brass gauze (about 10 mesh to the centimetre) may be soldered in the lid and in the bottom. Soft wheat of about 11% water-content is suitable for food. The can is about two-thirds filled with wheat; a hundred or so adults, or a handful of wheat from an established colony are added and the can is stored at 27°C. The stock may be maintained indefinitely if it is sub-cultured about every 2 months. It is convenient to accumulate 4 or 5 colonies and then, as each new one is started, the oldest is discarded. For the large 'strain' of *C. oryzae* the same methods may be used except that this species goes better on maize than on wheat.

Tribolium confusum and *T. castaneum*

Glass jars with a capacity of 1 to 1·5 litres are quite good. There is no need for covers if they are not filled beyond about three-quarters. The medium is made by mixing about 10% powdered yeast with white flour. Temperature and humidity are not critical but the beetles thrive best at a relative humidity of about 60% and at temperatures between 27° and 30°C. Begin by putting 100 adults into about 200 gm. of medium. After a week sieve out the adults and place them in a fresh jar. Return the egg-laden flour to the same jar and add about 100 gm. of medium. Repeat this each week until adults begin to appear in the first jar. Then as each new colony is founded throw the oldest one away. At 27°C this usually results in the maintenance of 6 or 8 colonies.

Tenebrio molitor

Wooden boxes with sliding lids, measuring about 40 × 40 × 20 cm., are suitable containers for rearing *Tenebrio*. The medium is a mixture, in equal parts, of bran, pollard and wheat germ. Place about 10 layers of jute sackcloth in the box and place several handfuls of medium between each layer. Add 100 adults and place the box at about 27°C. If a few pieces of fresh carrot or banana are kept in the box for the adults they are likely to lay more eggs. At 27°C the life-cycle occupies several months and at lower temperatures much longer. So it is desirable to maintain a number of colonies in different stages.

Mormoniella vitripennis

This species will breed on the pupae of many muscoid Diptera. We use blowflies, either *Calliphora* spp. or *Lucilia sericata*. The blowflies are bred by exposing meat in the open in tin cans with a little sawdust in the bottom. When sufficient eggs have been laid the cans are covered to keep flies out and the maggots are reared, adding more meat if necessary. In due course the pupae are collected from the sawdust.

The adults of *Mormoniella* feed on fluid which exudes from the blowfly pupa after it has been pierced by the wasp's ovipositor. The adults live for about 2 weeks at 25°C and lay several hundred eggs. At 25°C the life-cycle of *Mormoniella* takes about 14 days. But it is possible to induce a diapause in the fully-grown larvae by keeping the ovipositing females at 15°–18°C. Larvae in diapause may be stored at 5°C for upward of a year. But they will resume development in the warmth any time after they have been at 5°C for 3 months. This makes *Mormoniella* a very convenient animal to maintain in the laboratory.

Bibliography and Author Index

(*Numbers in italics denote references to text*)

AGASSI, J. (1959) 'Corroboration *versus* induction', *British J. Phil. Sci.*, **9:** 311–17. *180*

ALLEE, W. C., EMERSON, A. E., PARK, O., PARK, T., and SCHMIDT, K. P. (1949) *Principles of animal ecology*, Philadelphia: W. B. Saunders and Co. *3, 13, 78, 108*

ANDREWARTHA, H. G. (1943) 'The significance of grasshoppers in some aspects of soil conservation in South Australia and Western Australia', *J. Dept. Agric. South Australia*, **46:** 314–22. *154*

(1944) 'The distribution of plagues of *Austroicetes cruciata* Sauss (Acrididae) in Australia in relation to climate, vegetation and soil', *Tr. Roy. Soc. South Australia*, **68:** 315–26. *6, 154*

(1952) 'Diapause in relation to the ecology of insects', *Biol. Rev.*, **27:** 50–107. *70, 102*

(1957) 'The use of conceptual models in population ecology', *Cold Spring Harb. Symp. Quant. Biol.*, **22:** 219–36. *145, 147, 169, 173, 182*

(1959) 'Density-dependent factors in ecology', *Nature*, **183:** 200. *169, 182*

ANDREWARTHA, H. G., and BIRCH, L. C. (1954) *The distribution and abundance of animals*, Chicago: University of Chicago Press. *5, 7, 17, 42, 63, 104, 106, 108, 115, 125, 135, 156, 175, 181*

ANDREWARTHA, H. G., and BROWNING, T. O. (1958) 'Williamson's theory of interspecific competition', *Nature*, **181:** 1415. *171*

(1961) 'An analysis of the idea of "resources" in animal ecology', *J. Theor. Biol.* **1:** (*in press*). *146, 147*

BAILEY, N. T. J. (1951) 'On estimating the size of mobile populations from recapture data', *Biometrika*, **38:** 293–306. *43*

(1952) 'Improvements in the interpretation of recapture data', *J. Anim. Ecol.*, **21:** 120–7. *43*

BAKER, J. R., and RANSOM, R. M. (1932) 'Factors affecting the breeding of the field mouse (*Microtus agrestis*). I. Light', *Proc. Roy. Soc. London, B*, **110:** 313–21. *102*

BANKS, C. J. (1954) 'A method of estimating populations and counting large numbers of *Aphis fabae* Scop.', *Bull. Entom. Research*, **45:** 751–6. *20*

BARBER, G. W. (1939) 'Injury to sweet corn by *Euxesta stigmatias* Loew. in Florida', *J. Econ. Ent.*, **32:** 879–80. *50*

BATEMAN, M. A. (1955) 'The effect of light and temperature on the rhythm of pupal ecdysis in the Queensland fruit-fly *Dacus* (*Strumeta*) *tryoni* (Frogg.)', *Aust. J. Zool.*, **3**: 22–33. *98*

BEALL, G. (1946) 'Seasonal variation in sex proportion and wing-length in the migrant butterfly, *Danau plexippus* L. (Lep. Danaidae)', *Tr. Roy. Ent. Soc. London*, **97**: 337–53. *52*

BENTON, A. H., and WERNER, W. E. JR. (1958) *Principles of field biology and ecology*, New York, N.Y.: McGraw-Hill. *111*

BIRCH, L. C. (1953a) 'Experimental background to the study of the distribution and abundance of insects. I. The influence of temperature, moisture and food on the innate capacity for increase of three grain beetles', *Ecology*, **34**: 698–711. *14*, *115*

(1953b) 'Experimental background to the study of the distribution and abundance of insects. II. The relation between innate capacity to increase in numbers and the abundance of three grain beetles in experimental populations', *Ecology*, **37**: 712–26. *14*

(1953c) 'Experimental study to the distribution and abundance of insects. III. Relation between innate capacity for increase and survival of different species of beetles living together on the same food', *Evolution*, **7**: 136–44. *14*

(1957) 'The meanings of competition', *Amer. Nat.*, **91**: 5–18. *174*

BIRCH, L. C., and ANDREWARTHA, H. G. (1941) 'The influence of weather on grasshopper plagues in South Australia', *J. Dept. Agric. South Australia*, **45**: 95–100. *104*, *125*

BISSONNETTE, T. H. (1935) 'Modification of mammalian sex cycles. III. Reversal of the cycle in male ferrets (*Putorius vulgaris*) by increasing periods of exposure to light between October second and March thirtieth', *J. Exp. Zool.*, **71**: 341–67. *102*

BLAIR, F. (1951) 'Population structure, social behaviour, and environmental relations in a natural population of the beach mouse (*Peromyscus polionotus leucocephalus*)', *Contrib. Lab. Vert. Biol.* No. 48. *110*

BLISS, C. I. (1953) 'Fitting the negative binomial distribution to biological data and a note on the efficient fitting of the negative binomial', *Biometrics*, **9**: 176–200. *38*, *41*, *43*

BOGUSLAVSKY, G. W. (1956) 'Statistical estimation of the size of a small population', *Science*, **124**: 317–18. *34*

BRAIDWOOD, R. J., and REED, C. A. (1957) 'The achievement and early consequences of food-production: a consideration of the archeological and natural-history evidence', *Symposia Quant. Biol.*, **22**: 19–31. *139*

BRETT, J. R. (1944) 'Some lethal temperature relations of Algonquin Park fishes', *Univ. Toronto Stud. Biol.*, **52**. *71*

BRIAN, M. V. (1952) 'The structure of a natural dense ant population', *J. Anim. Ecol.*, **21**: 12–24. *176*

BRONOWSKI, J. (1951) *The commonsense of science*, London: Heinemann. *183*

BROWNING, L. O. (1952) 'The influence of temperature on the rate of development of insects, with special reference to the eggs of *Gryllulus commodus* Walker', *Australian J. Scient. Res., B*, **5:** 96–111. *229*

(1954) 'Water balance in the tick *Ornithodorus moubata* Murray, with particular reference to the influence of carbon dioxide on the uptake and loss of water', *J. Exp. Biol.*, **31:** 331–40. *91*

(1959) 'The long-tailed mealybug, *Pseudococcus adonidum* L. in South Australia', *Australian J. Agric. Res.*, **10:** 322–39. *19*

BRUES, C. T. (1939) 'Studies on the fauna of some thermal springs in Dutch East Indies', *Proc. Amer. Acad. Arts & Sc.*, **73:** 71–95. *78*

BULLOUGH, W. S. (1951) *Vertebrate sexual cycles*, London: Methuen and Co. *102*

BURNETT, T. (1949) 'The effect of temperature on an insect host-parasite population', *Ecology*, **30:** 113–34. *118*

BUXTON, P. A. (1955) *The natural history of tse-tse flies*, London: H. K. Lewis. *173*

CAMERON, J. McB. (1950) 'The laboratory of insect pathology', Sault Ste. Marie, Ontario, Forest Insects Investigations, *Bi-monthly Progress Rep.*, **6**, No. 4, 1–7. *131*

CARRICK, R. (1959) 'The contribution of banding to Australian bird ecology,' *Monograph. Biologicae*, **8:** 369–81. *21*, *108*

CARRICKER, M. R. (1951) 'Ecological observations on the distribution of oyster larvae in New Jersey estuaries', *Ecol. Monographs*, **21:** 19–38. *51*

CHITTY, D. (1952) 'Mortality among voles (*Microtus agrestis*) at Lake Vyrnwy, Montgomeryshire in 1936–9', *Phil. Trans. Roy. Soc. London, B*, **236:** 505–52. *112*

(1957) 'Self-regulation in numbers through changes in viability', *Symposia Quant. Biol.*, **22:** 277–80. *112*, *113*

CLARK, L. R. (1947) 'An ecological study of the Australian plague locust *Chortoicetes terminifera* Walk. in the Bogan-Macquarie outbreak area in N.S.W.', *Bull. Counc. Sci. Indust. Res. Australia*, No. 226. *72*

(1949) 'Behaviour of swarms of hoppers of the Australian plague locust *Chortoicetes terminifera* Walk.', *Bull. Counc. Sci. Indust. Res. Australia*, No. 245. *72*

CLARKE, G. L. (1954) *Elements of ecology*, New York: Wiley. *13*, *51*

CLAUSEN, C. P. (1951) 'The time factor in biological control', *J. Econ. Ent.*, **44:** 1–9. *118*

COAD, B. R. (1931) 'Insects captured by airplane are found at surprising heights', *Yearbook U.S. Dept. Agr.*, 1931: 320–3. *49*

COCKERELL, T. D. A. (1934) 'Mimicry' among insects, *Nature*, **133:** 329–330. *56*, *175*

COLE, L. C. (1946) 'A theory for analysing contagiously distributed populations', *Ecology*, **27:** 329–41. *40*

CRAGG, J. B. (1955) 'The natural history of sheep blowflies in Britain', *Ann. Appl. Biol.*, **42:** 197–207. *105*

CRIDDLE, N. (1930) 'Some natural factors governing the fluctuations of grouse in Manitoba', *Canadian Field Nat.*, **44:** 77–80. *138*

CROGHAN, P. C. (1958a) 'The survival of *Artemia salina* (L.) in various media', *J. Exp. Biol.*, **35:** 213–18. *80*

(1958b) 'The osmotic and ionic regulation in *Artemia salina* (L.)', *J. Exp. Biol.*, **35:** 219–33. *97*

(1958c) 'The mechanism of osmotic regulation in *Artemia salina* (L.); the physiology of the branchiae', *J. Exp. Biol.*, **35:** 234–42. *80*

(1958d) 'The mechanism of osmotic regulation in *Artemia salina* (L.); the physiology of the gut', *J. Exp. Biol.*, **35:** 243–9. *80*

DANTZIG, T. (1942) *Number the language of science*, New York: Scribner. *183*

DARLINGTON, P. J. (1938) 'The origin of the fauna of the Greater Antilles with a discussion of the dispersal of animals over water and through the air', *Quart. Rev. Biol.*, **13:** 274–300. *59*

(1957) *Zoogeography: the geographical distribution of animals*, New York: Wiley. *59*

DAVIDSON, J. (1944) 'On the relationship between temperature and the rate of development of insects at constant temperatures', *J. Anim. Ecol.*, **13:** 26–38. *229*

DAVIDSON, J., and ANDREWARTHA, H. G. (1948a) 'Annual trends in a natural population of *Thrips imaginis* (Thysanoptera)', *J. Anim. Ecol.*, **17:** 193–9. *74, 138*

(1948b) 'The influence of rainfall, evaporation and atmospheric temperature on fluctuations in size of a natural population of *Thrips imaginis* (Thysanoptera)', *J. Anim. Ecol.*, **17:** 200–2. *74, 138*

DAVIES, W. M. (1935) 'Studies on aphides infesting the potato crop. III. Effect of variation in relative humidity on the flight of *Myzus persicae* Sulz.', *Ann. Appl. Biol.*, **22:** 106–15. *48*

(1936) 'Studies on the aphides infesting the potato crop. V. Laboratory experiments on the effect of wind velocity on the flight of *Myzus persicae* Sulz.', *Ann. Appl. Biol.*, **23:** 401–8. *48*

(1939) 'Studies on aphides infesting the potato crop. VII. Report on a survey of the aphis population of potatoes in selected districts of Scotland (25 July–6 August 1936)', *Ann. Appl. Biol.*, **26:** 116–34. *48*

DAWSON, J. (1956) 'Splenic hypertrophy in voles', *Nature*, **178:** 1183–4. *113*

DE BACH, P. (1949) 'Population studies of the long-tailed mealybug and its natural enemies on citrus trees in southern California', *Ecology*, **30:** 14–25. *120*

DEEVEY, E. S. JR. (1956) 'The human crop', *Scient. Amer.*, **194:** 105–12. *139*

DICE, L. R. (1952) *Natural communities*, Ann. Arbor: University of Michigan Press. *13*

DICKSON, R. C. (1949) 'Factors governing the induction of diapause in the oriental fruit moth', *Ann. Entom. Soc. Amer.*, **42:** 511–37. *100*

DILL, D. B. (1938) *Life, heat and altitude*, Cambridge, Mass.: Harvard Univ. Press. *87*

DOBZHANSKY, TH., and WRIGHT, S. (1943) 'Genetics of natural populations. X. Dispersion rates in *Drosophila pseudoobscura*', *Genetics*, **28:** 304–40. *52*

(1947) 'Genetics of natural populations. XV. Rate of diffusion of a mutant gene through a population of *Drosophila pseudoobscura*', *Genetics*, **32:** 303–24. *52, 54, 210, 217*

DODD, A. P. (1936) 'The control and eradication of prickly pear in Australia', *Bull. Ent. Research*, **27:** 503–17. *145*

DORST, H. E., and DAVIS, E. W. (1937) 'Tracing long distance movements of the beet leafhopper in the desert', *J. Econ. Ent.*, **30:** 948–954. *50*

DOUDOROFF, P. (1938) 'Reactions of marine fishes to temperature gradients', *Biol. Bull.*, **75:** 494–509. *72*

DOWDESWELL, W. H. (1952) *Animal ecology*, London: Methuen. *13*

DOWDESWELL, W. H., FISHER, R. A., and FORD, E. B. (1940) 'The quantitative study of populations in the Lepidoptera'. I. *Polyommatus icarus*, *Ann. Eugenics*, **10:** 123–36. *25, 29*

DUFFY, E. (1956) 'Aerial dispersal in a known spider population', *J. Anim. Ecol.*, **25:** 85–111. *49*

DYMOND, J. R. (1947) 'Fluctuations in animal populations with special reference to those of Canada', *Trans. Roy. Soc. Canada*, **41:** 1–34. *115, 155*

EDNEY, E. B. (1947) 'Laboratory studies on the bionomics of the rat fleas *Xenopsylla brasiliensis* Baker and *X. cheopis* Roths. II. Water relations during the cocoon period', *Bull. Entom. Research*, **38:** 263–280. *97*

ELTON, C. (1927) *Animal ecology*, London: Sidgwick and Jackson. *3, 4, 9, 13, 57*

(1938) 'Animal numbers and adaptation'. In Beer, G. R. de (ed.), *Evolution: essays on aspects of evolutionary biology*, pp. 127–37, Oxford: University Press. *175*

(1949) 'Population interspersion: an essay on animal community patterns,' *J. Ecol.*, **37:** 1–23. *4, 13, 34, 155, 156*

(1958) *The ecology of invasions*, London: Methuen. *56, 59, 176*

ELTON, C., and MILLER, R. S. (1954) 'An ecological survey of animal communities: with a practical system of classifying habitats by structural characters', *J. Ecol.*, **42:** 460–96. *13, 155*

ERRINGTON, P. L. (1943) 'An analysis of mink predation upon muskrats in north-central United States', *Research Bull. Iowa Agr. Exp. Sta.*, **320:** 797–924. *21, 112, 122*

(1944) 'Ecology of the muskrat', *Rep. Iowa Agr. Exp. Sta.*, 1944: 187–189. *155*

(1945) 'Some contributions of a fifteen-year local study of the northern bobwhite to a knowledge of population phenomena', *Ecol. Monogr.*, **15:** 1–34. *21*

(1946) 'Predation and vertebrate populations', *Quart. Rev. Biol.*, **21:** 145–77 and 221–45. *21, 123*

(1948) 'Environmental control for increasing muskrat production', *Trans. 13th North Am. Wildlife Conference:* 596–609. *155*

ERRINGTON, P. L., and HAMMERSTROM, F. N. (1936) 'The northern bob-whites winter territory', *Research Bull. Iowa Agr. Exp. Sta.*, No. 201. *154*

EVANS, F. C. (1949) 'A population study of the house mouse (*Mus musculus*) following a period of local abundance', *J. Mammal.*, **30:** 351–63. *25*

EVANS, J. W. (1933) 'Thrips investigation I: The seasonal fluctuations in numbers of *Thrips imaginis* Bagnall and associated blossom thrips', *J. Counc. Sci. Ind. Res.* (Hust.), **6:** 145–59. *75*

FELT, E. P. (1925) 'The dissemination of insects by air currents', *J. Econ. Ent.*, **18:** 152–6. *50*

FINNEY, D. J. (1947) *Probit analysis: a statistical treatment of the sigmoid response curve*, Cambridge: at the University Press. *65*, *229*

FISHER, R. A. (1947) *The design of experiments* (*Fourth edition*), Edinburgh: Oliver and Boyd. *180*, *182*

FISHER, R. A., and YATES, F. (1948) *Statistical tables for biological, agricultural, and medical research workers*, London: Oliver and Boyd. *65*

FLANDERS, S. E. (1947) 'Elements of host discovery exemplified by parasitic Hymenoptera', *Ecology*, **28:** 299–309. *56*

FORD, E. B. (1945) *Butterflies*, London: Collins. *10*, *148*

FRAENKEL, G., and BLEWETT, M. (1944) 'The utilization of metabolic water in insects', *Bull. Entom. Research*, **35:** 127–37. *80*, *92*

FRAENKEL, G., and GUNN, D. L. (1940) *The orientation of animals: kineses, taxes and compass reactions*, Oxford: Clarendon Press. *86*

FRY, F. E. J. (1947) 'Effects of the environment on animal activity', *Univ. Tor. Stud. Biol.*, **55:** 1–62. *71*, *73*

FRY, F. E. J., BRETT, J. R., and CLAUSEN, G. H. (1942) 'Lethal limits of temperature for young goldfish', *Rev. Canad. de Biol.*, **1:** 50–6. *71*

FRY, F. E. J., HART, J. S., and WALKER, K. F. (1946) 'Lethal temperature relations for a sample of young speckled trout *Salvelinus fontinalis*', *Pub. Ontario Fish Research Lab.*, **66:** 5–35. *71*

GAUSE, G. F. (1934) *The struggle for existence*, Baltimore: Williams and Wilkins. *126*

GERKING, S. D. (1952) 'Vital statistics of the fish population of Gordy Lake, Indiana', *Trans. Amer. Fish. Soc.*, **82:** 48–67. *34*

GLOVER, P. E., JACKSON, C. H. N., ROBERTSON, A. G., and THOMSON, W. E. F. (1955) 'The extermination of the tse-tse fly, *Glossina morsitans* Westw., at Abercorn, Northern Rhodesia', *Bull. Ent. Res.*, **46:** 57–67. *148*

GOWERS, E. (1948) *Plain words*, London: H.M. Stat. Office. *12*

HALL, F. G. (1922) 'Vital limits of exsiccation of certain animals', *Biol. Bull.*, **42:** 31–51. *81*

HARDIN, G. (1956) 'Meaninglessness of the word protoplasm', *Sci. Monthly*, **82:** 112–20. *174*

HARKER, J. E. (1956) 'Factors controlling the diurnal rhythm of activity of *Periplaneta americana* L.', *J. Exp. Biol.*, **33:** 224–34. *98*

HOCHBAUM, H. A. (1955) *Travels and traditions of waterfowl*, Minneapolis: Univ. Minnesota Press. *52*

HOWARD, L. O., and FISKE, W. F. (1911) 'The importation into the United

States of the parasites of the gipsy moth and the brown-tail moth', *Bull. U.S. Bur. Entom.* No. 91. *169, 183*

HOY, J. M. (1954) 'A new species of *Eriococcus* Targ. (Hemiptera, Coccidae) attacking *Leptospermum* in New Zealand', *Trans. Roy. Soc. New Zealand,* **82:** 465–74. *49*

HUFFAKER, C. B. (1957) 'Fundamentals of the biological control of weeds', *Hilgardia,* **27:** 101–57. *183*

HUNTSMAN, A. G. (1946) 'Heat stroke in Canadian maritime stream fishes', *J. Fish Research Board Canada,* **6:** 476–82. *71*

HUTCHINSON, G. E., and DEEVEY, E. S. JR. (1949) 'Ecological studies on populations', *Survey Biol. Prog.,* **1:** 325–59. *105*

JACKSON, C. H. N. (1937) 'Some new methods in the study of *Glossina morsitans*', *Proc. Zool. Soc. London,* 1936: 811–96. *136*

(1939) 'The analysis of an animal population', *J. Anim. Ecol.,* **8:** 238–46. *25, 26, 31*

(1940) 'The analysis of a tse-tse fly population', *Ann. Eugenics,* **10:** 322–69. *43*

(1944) 'The analysis of a tse-tse fly population II', *Ann. Eugenics,* **12:** 176–205. *43*

(1948) 'The analysis of a tse-tse fly population III', *Ann. Eugenics,* **14:** 91–108. *43*

JOHNSON, B. (1953) 'Flight muscle autolysis and reproduction in aphids', *Nature,* **172:** 813. *45*

JOHNSON, C. G. (1940) 'The longevity of the fasting bed-bug (*Cimex lectularius*) under experimental conditions and particularly in relation to the saturation deficiency law of water loss', *Parasitology,* **32:** 329–70. *65, 66*

(1950a) 'The comparison of suction trap, sticky trap and tow net for the quantitative sampling of small airborne insects', *Ann. Appl. Biol.* **37:** 268–85. *20*

(1950b) 'Infestation of a bean field by *Aphis fabae* Scop. in relation to wind direction', *Ann. Appl. Biol.,* **37:** 441–50. *44*

(1952a) 'The changing numbers of *Aphis fabae* Scop. flying at crop level, in relation to current weather and to the population on the crop', *Ann. Appl. Biol.,* **39:** 525–47. *20, 44*

(1952b) 'A new approach to the problems of the spread of aphids and to insect trapping', *Nature,* **170:** 147–8. *20, 44*

(1954) 'Aphid migration in relation to weather', *Biol. Rev.,* **29:** 87–118. *44, 48*

(1955) 'Ecological aspects of aphid flight and dispersal', *Rep. Rothamsted Exp. Sta.,* 1955: 191–201. *20, 44, 47*

(1956) 'Changing views on aphid dispersal', *N.A.A.S. Quart. Rev.,* No. 32. *44*

(1957) 'The vertical distribution of aphids in the air and the temperature lapse rate', *Quart. J. Roy. Met. Soc.,* **83:** 194–201. *44*

JOHNSON, C. G., and TAYLOR, L. R. (1957) 'Periodism and energy summation with special reference to flight rhythms in aphids', *J. Exp. Biol.,* **34:** 209–21. *44*

JONES, E. P. (1937) 'The egg parasites of the cotton bollworm, *Heliothis armigera* Hubn. (*obsoleta Fubr.*) in Southern Rhodesia', *Pub. Brit. Sth Afr. Co.*, **6:** 37–105. *142*

KLUIJVER, H. N. (1951) 'The population ecology of the great tit *Parus m. major*', *Ardea*, **39:** 1–135. *21, 109, 157*

KOCH, H. (1938) 'The absorption of chloride ions by the anal papillae of diptera larvae', *J. Exp. Biol.*, **15:** 152–60. *89*

KROGH, A. (1937) 'Osmotic regulation in fresh water fishes by active absorption of chloride ions', *Z. Vergl. Physiol.*, **24:** 656–66. *89*

(1939) *Osmotic regulation in aquatic animals*, Cambridge: at the University Press. *79, 80, 88, 97, 102*

LACK, D. (1954) *The natural regulation of animal numbers*, London: Oxford University Press. *134*

LEEPER, G. W. (1941) 'The scientist's English', *Australian J. Sci.*, **3:** 121–123. *12*

(1952) 'An index of ease of reading', *Australian J. Sci.*, **15:** 31–2. *12*

LEES, A. D. (1946) 'The water balance in *Ixodes ricinus* L. and certain other species of ticks', *Parasitology*, **37:** 1–20. *91*

(1947) 'Transpiration and the structure of the epicuticle in ticks', *J. Exp. Biol.*, **23:** 379–410. *91, 92, 97*

(1955) *The physiology of diapause in arthropods*, Cambridge: at the University Press. *102*

LEOPOLD, A. (1933) *Game management*, New York: Charles Scribners Sons. *105, 147*

(1943) 'Deer irruptions', *Wisconsin Conserv. Bull.*, August, 1943. *147*

LESLIE, P. H. (1952) 'The estimation of population parameters from data obtained by means of the capture-recapture method', *Biometrika*, **39:** 363–8. *43*

LESLIE, P. H., and CHITTY, D. (1951) 'The estimation of population parameters from data obtained from the capture-recapture method. I. The maximum likelihood equations for estimating death-rate', *Biometrika*, **38:** 269–92. *43*

LLOYD, L. (1941) 'The seasonal rhythm of a fly (*Spaniotoma minima*) and some theoretical considerations', *Trans. Roy. Soc. Trop, Med. and Hyg.*, **35:** 93–104. *23*

(1943) 'Materials for a study in animal competition. II. The fauna of the sewage beds. III. The seasonal rhythm of *Psychoda alternata* Say, and an effect of intraspecific competition', *Ann. Appl. Biol.*, **30:** 47–60. *127*

LLOYD, L., GRAHAM, J. F., and REYNOLDSON, T. B. (1940) 'Materials for study, in animal competition. The fauna of the sewage bacteria beds', *Ann. Appl. Biol.*, **27:** 122–50. *127*

LOTKA, A. J. (1925) *Elements of physical biology*, Baltimore: Williams and Wilkins. *116*

LUDWIG, D., and CABLE, R. M. (1933) 'The effect of alternating temperatures on the pupal development of *Drosophila melanogaster*, Meigen', *Physiol. Zool.*, **6:** 493–508. *63*

McCABE, T. T., and BLANCHARD, B. D. (1950) Three species of *Peromyscus*, Santa Barbara, California: Rood. *150*

McDOUGALL, K. D. (1943) 'Sessile marine invertebrates of Beaufort, North Carolina', *Ecol. Monogr.*, **13:** 323–74. *106*

MACFADYEN, A. (1957) *Animal ecology: aims and methods*, London: Pitman. *13*

MACKENZIE, J. M. D. (1952) 'Fluctuations in the numbers of British tetraonids', *J. Anim. Ecol.*, **21:** 128–53. *15*

MACKERRAS, I. M., and MACKERRAS, M. J. (1944) 'Sheep blowfly investigations. The attractiveness of sheep for *Lucilia sericata*', *Bull. Counc. Sci. Indust. Research Australia*, No. 181. *127*

MADGE, P. E. (1954) 'Control of the underground grass caterpillar', *J. Dept. Agric. S. Australia*, March 1954. *23*

MAELZER, D. A. (1958) 'The ecology of *Aphodius howitti*', Thesis (unpublished) in library of University of Adelaide. *85*

MAIL, G. A., and SALT, R. W. (1933) 'Temperature as a possible limiting factor in the northern spread of the Colorado potato beetle', *J. Econ. Ent.*, **26:** 1068–75. *151*

MARSHALL, F. H. A. (1942) 'Exteroceptive factors in sexual periodicity', *Biol. Rev.*, **17:** 68–89. *102*

MATHER, K. (1943) *Statistical analysis in Biology*, London: Methuen. *18*

MATHEWS, G. V. T. (1955) *Bird navigation*, Cambridge: at the University Press. *52*

MATTHÉE, J. J. (1951) 'The structure and physiology of the egg of *Locustana pardalina* (Walk)', *Scient. Bull. Dept. Agr. Sth. Africa*, No. 316. *90*

MAUGHAM, W. S. (1948) *The summing up*, London: Heinemann. *11*

MILNE, A. (1943) 'The comparison of sheep tick populations (*Ixodes ricinus* L.)', *Ann. Appl. Biol.*, **30:** 240–50. *23*

(1949) 'The ecology of the sheep tick *Ixodes ricinus* L. Host relationships of the tick. 2. Observations on hill and moorland grazings in northern England', *Parasitology*, **39:** 173–97. *103, 135*

(1950) 'The ecology of the sheep tick *Ixodes ricinus* L. Spatial distribution, *Parasitology*', **40:** 35–45. *103*

(1951) 'The seasonal and diurnal activities of individual sheep ticks (*Ixodes ricinus* L.)', *Parasitology*, **41:** 189–208. *103, 135*

(1952) 'Features of the ecology and control of the sheep tick, *Ixodes ricinus* L. in Britain', *Ann. Appl. Biol.*, **39:** 144–6. *103*

(1957) 'Theories of natural control of insect populations', *Cold Spring Harbour, Symp. Quant. Biol.*, **22:** 253–71. *168, 171*

MOORE, J. A. (1939) 'Temperature tolerance and rates of development in the eggs of Amphibia', *Ecology*, **20:** 459–78. *77*

(1942) 'The role of temperature in the speciation of frogs', *Biol. Symp.*, **6:** 189–213. *77*

MORONEY, M. J. (1953) *Facts from figures*, London: Penguin Books. *18, 221*

NICHOLSON, A. J. (1933) 'The balance of animal populations', *J. Anim. Ecol.*, **2:** 132–78. *168, 170, 172, 182*

(1937) 'The role of competition in determining animal populations', *J. Counc. Sci. Indust. Research, Australia*, **10:** 101–6. *172*

NICHOLSON, A. J. (1947) 'Fluctuation of animal populations', *Rep. 26th Meeting A.N.Z.A.A.S.*, Perth, 1947. *58, 145, 174*

(1950) 'Population oscillations caused by competition for food', *Nature*, **165:** 476–7. *126, 141*

(1954) 'An outline of the dynamics of animal populations', *Australian J. Zool.*, **2:** 9–65. *168, 170, 172, 182*

(1958) 'Dynamics of insect populations', *Ann. Rev. Entom.*, **3:** 107–36. *169, 172, 174*

ODUM, E. P. (1953) *Fundamentals of ecology*, Philadelphia: Saunders. *13*

ORWELL, G. (1946) 'The English language', *J. and Proc. Aust. Chem. Inst.*, Sept. 1946. *13*

PARK, T. (1948) 'Experimental studies of interspecies competition. I. Competition between populations of flour beetles *Tribolium confusum* Duval and *Tribolium castaneum* Herbst', *Ecol. Monog.*, **18:** 265–308. *107, 126*

(1954) 'Experimental studies of interspecies competition. II. Temperature, humidity and competition in two species of *Tribolium*', *Physiol. Zool.*, **27:** 177–238. *107*

PAYNE, N. M. (1927) 'Measures of insect cold-hardiness', *Biol. Bull.*, **52:** 449–57. *81*

PEARSON, O. P. (1947) 'The rate of metabolism of some small mammals', *Ecology*, **28:** 127–45. *106*

PETRUSEWICZ, K. (1959) 'Investigations of experimentally induced population growth', *Ekolog. Polska Ser A*, **5:** 281–309. *116*

PITELKA, F. A. (1951) 'Ecologic overlap and interspecific strife in breeding populations of Anna and Allen humming birds', *Ecology*, **32:** 641–61. *176*

POLLISTER, A. W., and MOORE, J. A. (1937) 'Tables for the normal development of *Rana sylvatica*', *Anat. Rec.*, **68:** 489–96. *77*

POTTS, W. H., and JACKSON, C. H. N. (1953) 'The Shinyanga game-destruction experiment', *Bull. Entom. Res.*, **43:** 365–74. *136*

POWSNER, L. (1935) 'On the effects of temperature on the duration of the developmental stages of *Drosophila melanogaster*', *Physiol. Zool.*, **8:** 474–520. *61*

PREYER, W. (1891) 'Ueber die Anabiose', *Biol. Zentralbl.*, **11:** 1–5. *81*

RAINEY, R. C., and WALOFF, Z. (1948) 'Desert locust migrations and synoptic meteorology in the Gulf of Aden area', *J. Anim. Ecol.*, **17:** 110–12. *50.*

(1951) 'Flying locusts and convection currents', *Bull. Anti-loc. Res. Centre*, London, **9:** 51–70. *50*

RAMSAY, J. A. (1952) *Physiological approach to the lower animals*, Cambridge: at the University Press. *88*

RICHARDS, L. A. (1949) 'Methods of measuring soil moisture tension', *Soil Science*, **68:** 95–112. *85*

RICKER, W. E. (1954) 'Stock and recruitment', *J. Fish. Res. Board Canada*, **11:** 559–623. *116*

ROTH, L. M., and WILLIS, E. R. (1951) 'The effects of desiccation and starvation on the humidity behaviour and water balance of *Tribolium confusum* and *Tribolium castaneum*', *J. Exp. Zool.*, **118:** 337–61. *81, 82*

ROWAN, W. (1938) 'Light and seasonal reproduction in animals', *Biol. Rev.*, **13:** 374–402. *101*

RUSSELL, B. (1956) *Portraits from memory*, London: George Allen and Unwin. *182*

RUSSELL, F. S. (1939) 'Hydrographical and biological conditions in the North Sea as indicated by Plankton organisms', *J. Cons. Int. Explor. Mer.*, **14:** 171–92. *51*

SALT, G., HOLLICK, F. S. J., RAW, F., and BRIAN, M. V. (1948) 'The arthropod population of pasture soil', *J. Anim. Ecol.*, **17:** 139–50. *22*

SALT, R. W. (1950) 'The time as a factor in the freezing of undercooled insects', *Canad. J. Research, D.*, **25:** 66–86. *67*

SCHMIDT, J. (1925) 'The breeding places of the eel', *Smithsonian Inst. Ann. Rep.*, 1924: 279–316. *51*

SCHMIDT, P. (1918) 'Anabiosis of the earthworm', *J. Exp. Zool.*, **27:** 57–72. *81*

SCHMIDT-NIELSEN, B., and SCHMIDT-NIELSEN, K. (1949) 'The water economy of desert mammals', *Sci. Monthly*, **69:** 180–5. *87*

(1950) 'Evaporative water loss in desert rodents in their natural habitat', *Ecology*, **31:** 75–85. *87*

(1951) 'A complete account of the water metabolism in the kangaroo rat and an experimental verification', *J. Cell. Comp. Physiol.*, **38:** 165–81. *87*

SCHMIDT-NIELSEN, B., SCHMIDT-NIELSEN, K., and JARNUM, S. A. (1956) 'Water balance of the camel', *Amer. J. Physiol.*, **185:** 185–94. *81, 97*

SCHMIDT-NIELSEN, K., and SCHMIDT-NIELSEN, B. (1952) 'Water metabolism of desert mammals', *Physiol. Rev.*, **32:** 135–66. *81, 93, 94, 97*

SCHMIDT-NIELSEN, K., and SLADEN, W. J. L. (1958) 'Nasal salt secretion in the Humboldt penguin', *Nature*, **181:** 1217–18. *97*

SHELFORD, V. E. (1927) 'An experimental study of the relations of the codling moth to weather and climate', *Bull. Illinois Nat. Hist. Surv.*, **16:** 311–440. *69*

SMITH, H. S. (1935) 'The rôle of biotic factors in determining population densities', *J. Econ. Entom.*, **28:** 873–98. *170, 175*

(1939) 'Insect populations in relation to biological control', *Ecol. Monog.*, **9:** 311–20. *57, 118*

SOLOMON, M. E. (1957) 'Dynamics of insect populations', *Ann. Rev. Entom.*, **2:** 121–42. *173*

SOUTHWICK, C. H. (1955a) 'The population dynamics of confined house mice supplied with unlimited food', *Ecology*, **36:** 212–25. *112, 113*

(1955b) 'Regulatory mechanisms of house mouse populations: social behaviour affecting litter survival', *Ecology*, **36:** 627–34. *112, 113*

STRECKER, R. L., and EMLEN, J. T. JR. (1953) 'Regulatory mechanisms in house-mouse populations: the effect of limited food supply on a confined population', *Ecology*, **34:** 375–85. *112, 113*

'STUDENT' (1919) 'An explanation of deviations from Poisson's law in practice', *Biometrika*, **12:** 211–15. *40*

TAYLOR, L. R. (1957) 'Temperature relations of teneral development and behaviour in *Aphis fabae* Scop.', *J. Exp. Biol.*, **34:** 189–208. *44, 45*

THORSON, T., and SVIHLA, A. (1943) 'Correlation of the habitats of amphibians with their ability to survive the loss of body water', *Ecology*, **24:** 374–81. *81*

ULLYETT, G. C. (1947) 'Mortality factors in populations of *Plutella maculipennis* Curtiss (Tineidae Lepidoptera) and their relation to the problem of control', *Mem. Dept. Agric. Sth. Africa*, **2:** 77–202. *129*

(1950) 'Competition for food and allied phenomena in sheep blowfly populations', *Phil. Trans. Roy. Soc. London, B*, **234:** 77–174. *140, 176*

VARLEY, G. C. (1947) 'The natural control of population balance in knapweed gall fly (*Urophora jaceana*)', *J. Anim. Ecol.*, **16:** 139–87. *170*

VOLTERRA, V. (1931) *Leçons sur la théorie mathématique de la lutte pour la vie* (Cahier Scient., Vol. 7), Paris: Cauthiers-Villars. *126*

WADDINGTON, C. H. (1956) 'Genetic assimilation of the bithorax phenotype', *Evolution*, **10:** 1–13. *181*

WALOFF, Z. (1946a) 'A long-range migration of the desert locust from southern Morocco to Portugal, with an analysis of concurrent weather conditions', *Proc. Roy. Ent. Soc. London, A*, **21:** 81–4. *50*

(1946b) 'Seasonal breeding and migrations of the desert locust (*Schistocerca gregaria* Forsk.) in eastern Africa', *Mem. Anti-locust Research Centre London*, No. 1. *50*

WATERHOUSE, D. F. (1947) 'The relative importance of live sheep and of carrion as breeding grounds for the Australian sheep blowfly *Lucilia cuprina*', *Bull. Counc. Sci. Indust. Research Australia* No. 217. *126*

WEISS, P. (1950) 'Perspectives in the field of morphogenesis', *Quart. Rev. Biol.*, **25:** 177–98. *12, 178*

WELLINGTON, W. G. (1949) 'The effects of temperature and moisture upon the behaviour of the spruce budworm, *Choristoneura fumiferana* Clemens (Lepidoptera: Tortricidae) I. The relative importance of graded temperatures and rates of evaporation in producing aggregations of larvae', *Scient. Agric.*, **29:** 201–51. *82*

WICHTERMAN, R. (1953) *The biology of* Paramoecium, New York: Blakiston. *212*

WIGGLESWORTH, V. B. (1932) 'On the function of the so called rectal glands of insects', *Quart. J. Micr. Sci.*, **75:** 131–50. *89*

(1933) 'The effect of salt on the anal gills of the mosquito larvae', *J. Exp. Biol.*, **10:** 1–15. *89*

(1948) 'The insect cuticle', *Biol. Rev.*, **23:** 408–51. *91*

WILLIAMS, C. B. (1940) 'An analysis of four years' captures of insects in a light trap. II. The effect of weather conditions on insect activity; and the estimation and forecasting of changes in the insect population', *Trans. Roy. Entom. Soc. London*, **90:** 227–306. *21*

WILLIAMSON, M. H. (1957) 'An elementary theory of interspecific competition', *Nature*, **180:** 422–4. *171*

Index

PHOENIX BOOKS
in Science

P 21 *W. E. LeGros Clark:* History of the Primates
P 35 *C. F. von Weizsäcker:* The History of Nature
P 40 *Giorgio de Santillana:* The Crime of Galileo
P 58 *Laura Fermi:* Atoms in the Family: Life with Enrico Fermi
P 126 *Alan Gregg:* For Future Doctors
P 155 *Michael Polanyi:* Science, Faith and Society
P 159 *Thomas S. Kuhn:* The Structure of Scientific Revolutions

PHOENIX SCIENCE SERIES

PSS 501 *Carey Croneis and William C. Krumbein:* Down to Earth: An Introduction to Geology
PSS 502 *Mayme I. Logsdon:* A Mathematician Explains
PSS 503 *Heinrich Klüver:* Behavior Mechanisms in Monkeys
PSS 504 *Gilbert A. Bliss:* Lectures on the Calculus of Variations
PSS 505 *Reginald J. Stephenson:* Exploring in Physics
PSS 506 *Harvey Brace Lemon:* From Galileo to the Nuclear Age
PSS 507 *A. Adrian Albert:* Fundamental Concepts of Higher Algebra
PSS 509 *Walter Bartky:* Highlights of Astronomy
PSS 510 *Frances W. Zweifel:* A Handbook of Biological Illustration
PSS 511 *Henry E. Sigerist:* Civilization and Disease
PSS 512 *Enrico Fermi:* Notes on Quantum Mechanics
PSS 513 *S. Chandrasekhar:* Plasma Physics
PSS 514 *A. A. Michelson:* Studies in Optics
PSS 515 *Gösta Ehrensvärd:* Life: Origin and Development
PSS 516 *Harold Oldroyd:* Insects and Their World
PSS 517 *A. Adrian Albert:* College Algebra
PSS 518 *Winthrop N. Kellogg:* Porpoises and Sonar
PSS 519 *H. G. Andrewartha:* Introduction to the Study of Animal Populations
PSS 520 *Otto Toeplitz:* The Calculus: A Genetic Approach
PSS 521 *David Greenhood:* Mapping
PSS 522 *Fritz Heide:* Meteorites
PSS 523 *Robert Andrews Millikan:* The Electron: Its Isolation and Measurements and the Determination of Some of Its Properties
PSS 524 *F. A. Hayek:* The Sensory Order
PSS 525 *A. Ya. Khinchin:* Continued Fractions

PHOENIX BOOKS

Sociology, Anthropology, and Archeology

P 2 *Edward Chiera:* They Wrote on Clay
P 7 *Louis Wirth:* The Ghetto
P 10 *Edwin H. Sutherland,* EDITOR: The Professional Thief, by a professional thief
P 11 *John A. Wilson:* The Culture of Ancient Egypt
P 20 *Kenneth P. Oakley:* Man the Tool-maker
P 21 *W. E. LeGros Clark:* History of the Primates
P 24 *B. A. Botkin,* EDITOR: Lay My Burden Down: A Folk History of Slavery
P 28 *David M. Potter:* People of Plenty: Economic Abundance and the American Character
P 31 *Peter H. Buck:* Vikings of the Pacific
P 32 *Diamond Jenness:* The People of the Twilight
P 45 *Weston La Barre:* The Human Animal
P 53 *Robert Redfield:* The Little Community *and* Peasant Society and Culture
P 55 *Julian A. Pitt-Rivers:* People of the Sierra
P 64 *Arnold van Gennep:* The Rites of Passage
P 71 *Nels Anderson:* The Hobo: The Sociology of the Homeless Man
P 82 *W. Lloyd Warner:* American Life: Dream and Reality
P 85 *William R. Bascom and Melville J. Herskovits,* EDITORS: Continuity and Change in African Cultures
P 86 *Robert Redfield and Alfonso Villa Rojas:* Chan Kom: A Maya Village
P 87 *Robert Redfield:* A Village That Chose Progress: Chan Kom Revisited
P 88 *Gordon R. Willey and Philip Phillips:* Method and Theory in American Archaeology
P 90 *Eric Wolf:* Sons of the Shaking Earth
P 92 *Joachim Wach:* Sociology of Religion
P 105 *Sol Tax,* EDITOR: Anthropology Today: Selections
P 108 *Horace Miner:* St. Denis: A French-Canadian Parish
P 117 *Herbert A. Thelen:* Dynamics of Groups at Work
P 124 *Margaret Mead and Martha Wolfenstein,* EDITORS: Childhood in Contemporary Cultures
P 125 *George Steindorff and Keith C. Seele:* When Egypt Ruled the East
P 129 *John P. Dean and Alex Rosen:* A Manual of Intergroup Relations
P 133 *Alexander Heidel:* The Babylonian Genesis
P 136 *Alexander Heidel:* The Gilgamesh Epic and Old Testament Parallels
P 138 *Frederic M. Thrasher:* The Gang: A Study of 1,313 Gangs in Chicago (Abridged)
P 139 *Everett C. Hughes:* French Canada in Transition
P 162 *Thomas F. O'Dea:* The Mormons
P 170 *Anselm Strauss,* EDITOR: George Herbert Mead on Social Psychology
P 171 *Otis Dudley Duncan,* EDITOR: William F. Ogburn on Culture and Social Change
P 172 *Albert J. Reiss, Jr.,* EDITOR: Louis Wirth on Cities and Social Life
P 173 *John F. Embree:* Suye Mura: A Japanese Village
P 174 *Morris Janowitz:* The Military in the Political Development of New Nations
P 175 *Edward B. Tylor:* Researches into the Early History of Mankind and the Development of Civilization
P 176 *James Mooney:* The Ghost-Dance Religion and the Sioux Outbreak of 1890